Computational Design

for Landscape Architects

This book is a guide to computational design for landscape architects replete with extensive tutorials. It introduces algorithmic approaches for modeling and designing landscapes. The aim of this book is to use algorithms to understand and design landscape as a generative system, i.e. to harness the processes that shape landscape to generate new forms. An algorithmic approach to design is gently introduced through visual programming with Grasshopper, before more advanced methods are taught in Python, a high-level programming language. Topics covered include parametric design, randomness and noise, waves and attractors, lidar, point cloud modeling, terrain modeling, earthworks, digital fabrication, and more. The chapters include sections on theory, methods, and either visual programming or scripting. Online resources for the book include code and datasets so that readers can easily follow along and try out the methods presented. This book is a much-needed guide, both theoretical and practical, on computational design for students, educators, and practitioners of landscape architecture.

Brendan Harmon is a computational designer, spatial scientist, and Assistant Professor of Landscape Architecture at Louisiana State University, where he teaches geographic information systems, computational design, digital fabrication, and robotics.

Computational Design

for Landscape Architects

Brendan Harmon

Routledge
Taylor & Francis Group
NEW YORK AND LONDON

Designed cover image: © Brendan Harmon

First published 2024
by Routledge
605 Third Avenue, New York, NY 10158

and by Routledge
4 Park Square, Milton Park, Abingdon, Oxon, OX14 4RN

Routledge is an imprint of the Taylor & Francis Group, an informa business

Library of Congress Cataloging-in-Publication Data
A catalog record for this title has been requested

ISBN: 9781032407043 (hbk)
ISBN: 9781032407050 (pbk)
ISBN: 9781003354376 (ebk)

DOI: 10.4324/9781003354376

Typeset in Plex
by Brendan Harmon

Publisher's Note:
This book has been prepared from camera-ready copy provided by the author.

Contents

Preface

This book is a practical guide to computational design for landscape architects. An algorithmic approach to design is gently introduced through visual programming with Grasshopper, before more advanced methods are taught in Python, a high-level programming language.

0.1 How to Read this Book

Grasshopper First-time readers should focus on the sections of the book covering visual programming with Grasshopper, skipping the more advanced sections on Python and Rhino Scripting. Consider skipping the chapter on programming entirely, but read the relevant sections of the appendices early on.

1 Introduction › 2 Visual Programming › A Packages › B Color › C File Paths › D Code & Data › 4 Randomness › 5 Random Paving › 6 Tessellations › 7 Point Clouds › 8 Grove › 9 Waves › 10 Surfaces › 11 Patterns › 12 Attractors › 13 Noise › 14 Biome › 15 Terrain › 16 Dunes › 17 Earthworks › 18 Fabrication

Python After reading and working your way through the Grasshopper sections of the book, try the Python sections. The chapter on programming introduces the basics of Python, while appendix A on packages is essential for getting started and should be read very early on.

A Packages › D Code & Data › 3 Programming › 4 Randomness › C File Paths › 7 Point Clouds › B Color › 9 Waves › 10 Surfaces › 13 Noise › 15 Terrain › 16 Dunes › 18 Fabrication

Rhino Scripting Read the sections on Rhino Scripting after mastering Grasshopper and Python.

C File Paths › D Code & Data › 3 Programming › 7 Point Clouds › 15 Terrain › 16 Dunes

0.2 Getting Started

Download all of the code and data for this book from `https://doi.org/10.5281/zenodo.8191264` and then continue to the introduction.

1 Introduction

Landscape is a dynamic medium and its form – the shape of the earth and the mass of plants – is continually shaped by natural processes such as the flow of water and sediment and the succession of ecological communities. Algorithms are used to describe and predict the physical processes that shape landscapes such as the erosive force of raindrops detaching sediment, the flow of water over the land transporting and depositing sediment, and the settling of deposited sediment under the force of gravity. This book is a guide to using algorithms – sequential sets of rules for performing tasks – to creatively and scientifically model landscapes. The aim of the book is to use algorithms to understand and design landscape as a generative system, i.e. to harness the processes that shape landscape to generate new forms. Algorithmic thinking not only underpins the paradigms of theoretical, computational, and data science, but also the creative practice of computational design. While spatial scientists use algorithms, implemented in code, to model, analyze, and simulate phenomena in space and time, designers use algorithms to generate, evaluate, and fabricate complex forms. Geomorphologists, for example, use algorithms to model the shape of the Earth's surface, to analyze its slope and landforms, and to simulate how water flows across it, eroding sediment and reshaping the land. Ecologists use algorithms to map species' habitats, model population dynamics, simulate migration routes, and design ecological corridors. Architects use algorithms creatively to model complex architectural forms and then digitally fabricate or robotically construct their designs. While algorithms have long been used by landscape architects to analyze landscapes, the discipline is beginning to use algorithms creatively to generate complex landforms, planting patterns, and circulation systems.

Figure 1.1: Benedikt Groß, *Avena+ Test Bed*, 2013

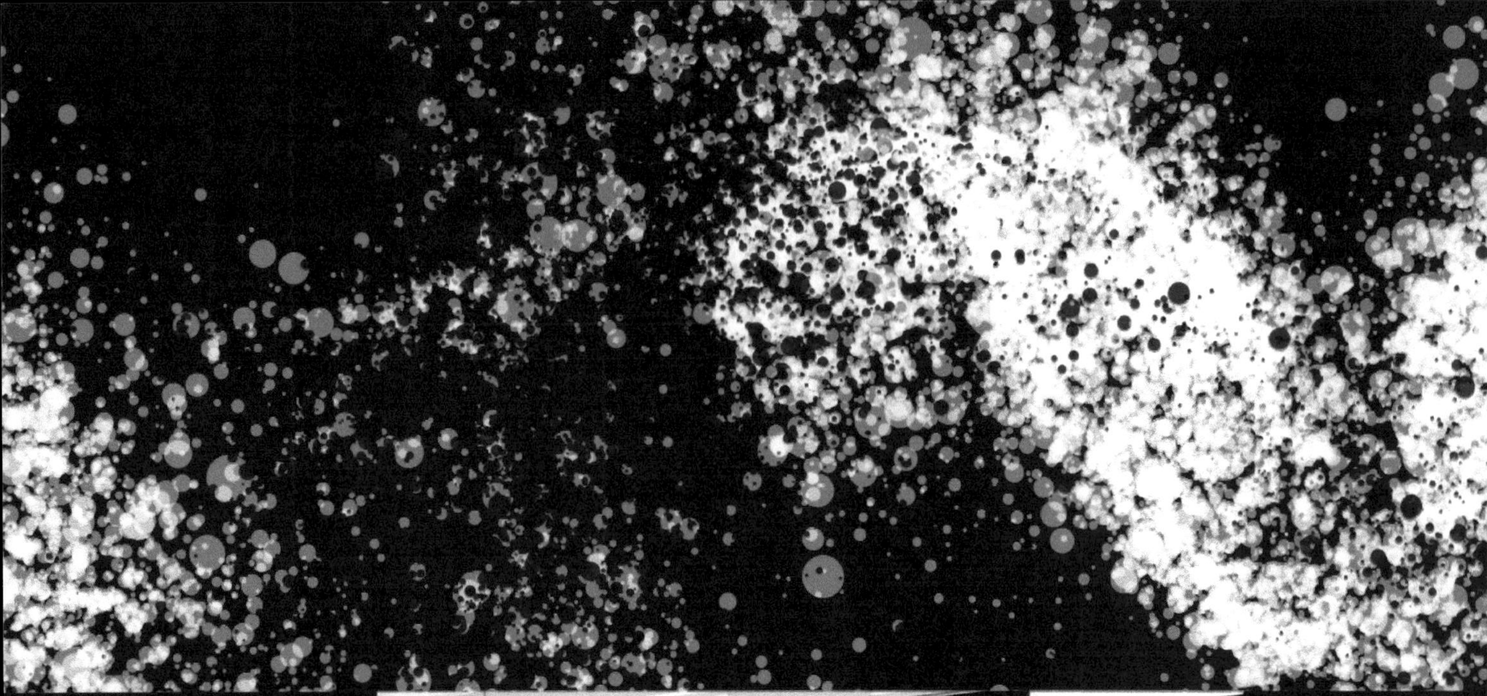

1.1 Algorithms

Algorithms – "precisely defined set[s] of mathematical or logical operations for the performance of a particular task" [3] – when used creatively in art and design evoke a peculiar aesthetics of infinity. The term algorithm is derived from a Latinization of Muhammad ibn Mûsâ al-Khwârizmî, a medieval Persian polymath and the father of algebra [2, 51]. Algorithms that model the behavior of natural systems, such as fractals, cellular automata, branching and venation, differential growth, hyphae, slime molds, and flocking, can produce infinite, complex patterns that evolve over time. In generative art, artists use such algorithms as performances that hide, reveal, and transform meaning. As the algorithm runs, step by step, the performance unfolds. Although the code of the algorithm may remain hidden, the artist's intention is transformed by the algorithm; the artwork's meaning accumulates through performance. Even when the audience only sees a final resulting image or form – rather than an ongoing animation or performance – the algorithm leaves its mark, traces of its process, of its logic. Algorithms can express infinity through endless performance and the emergence of complexity from simple rules. While works of generative art and design are diverse, spanning a wide range of disciplines, media, and themes, there is a common current – the aesthetic of the algorithm, an aesthetic of enfolding and unfolding [51]. Laura Mark's aesthetics of enfolding and unfolding is a semiotic process for digital media in which the infinite is unfolded through its transcription into information and then image. This algorithmic aesthetics of infinity is abstract, non-representational, and performative.

Figure 1.2: Jared S. Tarbell, *Infinite Regeneration*, 2021

Figure 1.3: Ryoji Ikeda, *test pattern [live set]*, 2010, photograph by Nacho Pintos

Figure 1.4: Ryoji Ikeda, *data.path*, 2013, photograph by r2hox

1.2 Generative Art

Works of generative art such as Jared S. Tarbell's Infinite Regeneration (Figure 1.2) [6], Ryoji Ikeda's test patterns (Figure 1.3) [52] and data.path (Figure 1.4) [44], Davide Quayola's Pleasant Places (Figure 1.5) [39] and Remains: Vallée de Joux (Figure 1.6) [23], and Rafael Lozano-Hemmer's Babbage Lovelace [15] use algorithms to evoke infinity and the emergence of complexity over time. Ryoji Ikeda's test patterns are a series of audiovisual installations in which his minimalist electronic music is rendered in real time as an animated matrix of black and white cells. This matrix representing the waveform of the song is projected onto the walls and floors of a dark space. The cells change rapidly, with time traced in rows like barcodes, flickering in response to the music. In this installation seemingly random bars of music and light challenge the audience to perceive a pattern, to experience the emergence of a system from binary interactions. In Quayola's Pleasant Places, part of the Landscape Paintings series, film footage of a landscape is gradually transformed by algorithms to reveal the sublime in nature. As image analysis, image processing, and audio signal processing algorithms progressively distort the original footage and sound, the landscape becomes more and more abstract with windswept leaves, grasses, and flowers gradually transforming into swirls of color. In its eventual abstraction this video alludes to the impressionist tradition of landscape painting, but as a performance it explores the dynamic nature of sensation, memory, and the environment. This algorithmic painting, as it transforms from a realistic scene to an abstract field of animate color, reveals primordial qualities of nature by questioning how we see and remember. Quayola's Remains are a series of point cloud renderings of woodland scenes, captured with laser scanning and rendered as points. By using lasers to capture these scenes, Quayola explores light as a medium. These point cloud renderings are both indexical representations of reality and algorithmic abstractions. The imperfections of the technology give rise to a painterly, pointillist aesthetic that evokes the material and immaterial.

Figure 1.5: Davide Quayola, *Pleasant Places*, 2015

Figure 1.6: Davide Quayola, *Remains: Vallée de Joux*, 2018

Figure 1.7: dECOi Architects, *One Main*, 2011

EXIT

1.3 Computational Design

In computational design this aesthetics of the algorithm is given form in landscape and the built environment. In computational design, designers, rather than drawing or modeling a singular form, write code or compose visual programs that generate form. The designer scripts a creative process, a generative system. By designing algorithms rather than discrete forms – whether art objects, products, buildings, or landscapes – designers can rapidly explore design ideas and generate variations on a theme. By generating variations on a theme, designers can create families of form where each design is unique, but related, formed by the same system, by the same logic. When confronted with endless variations on a theme, computational designers can use algorithms to analyze their designs and identify which perform best. With technologies for digital fabrication – such as computer numerical controlled (CNC) machining, 3D printing, and robotics – even complex algorithmically designed forms can be built precisely and efficiently. The use of algorithms to both design and then fabricate forms enables efficient non-standard production, i.e. mass customization, as an alternative to the mechanical mass production of identical copies [54, 48]. Entire structures or landforms can be autonomously constructed with robotic systems such as industrial robotic arms and autonomous earthmovers. Robotics promises to bridge the divide between design and construction [42], with the designer writing the algorithms that generate form and then build it.

1.3.1 Architecture

Rule-based design has a long history in architecture predating computers. Architectural rules can be expressed, not only by algorithm, but also by systems of proportions, sequences of geometric transformations, or physical experiments that put the laws of physics into practice. Medieval master builders, for example, followed geometric rules with the aid of tools and templates. Antoni Gaudí used generative systems such as hyperbolic paraboloids in Sagrada Familia [33], while Frei Otto often used physical models such as hanging chains and soap films to develop the form of his tensile structures [56]. Now designers use algorithms creatively to generate and transform geometry, incorporate randomness and noise, work with data, simulate physical processes, and interact with users or the environment. Architects have used algorithms to parametrically model geometry, generate families of elements [55], create non-standard building modules [24], design and test vaults [11], design complex façades [37, 20], and craft interactive experiences [32]. Similarly algorithms can be used by landscape architects to generate paving [29] and planting patterns, to shape topography [36] and hardscape, to model non-standard elements such as furniture, and to simulate processes such as the movement of people, the flow of water, the dispersal of seeds, or the succession of ecological communities.

Figure 1.8: MAEID, *Pahoehoe Beauty*, 2018

Figure 1.9: MAEID, *Magic Queen*, 2021

Figure 1.10: MAEID, *Sylva*, 2021, photograph by Zita Oberwalder

Digital Fabrication Digital fabrication, by giving architects more control over the construction of their designs, has spurred structural, material, and aesthetic explorations in architecture. Architects use technologies such as laser and plasma cutting, CNC routing and milling, 3D printing, and robotics [45] to build prefabricated parts for complex assemblies. Projects such as dECOi Architects' interior for One Main (Figure 1.7) [49] and Snøhetta's Norwegian Wild Reindeer Centre Pavilion [50] used CNC milling to carve unique wooden components for organic, parametrically modeled forms. The interior of One Main was assembled from non-standard, CNC milled panels of sustainably harvested spruce plywood to create a novel, sinuous aesthetic in which architectural elements and furniture blend together as a cohesive whole. Similarly, the organic form of Tverrfjellhytta, the Norwegian Wild Reindeer Centre Pavilion, was parametrically modeled and then assembled from non-standard, CNC milled beams of pine timber. While digital fabrication has extended architectural agency from the design office to the machine shop, robotics promises to extend architectural agency to the construction site.

Robotics Architects have been experimenting with robotic processes for autonomously constructing buildings such as laying bricks [40, 35, 28], stone cutting [33], assembling timber frames [26, 18], constructing complex formwork [34, 13], winding tensile structures [30, 12], and 3D printing metal [21], concrete [31], and soil structures [17, 9]. When robots, programmed by architects and engineers, construct buildings on site, the design team has direct control over construction. With robotic construction architecture becomes performative; architects are no longer just designing form, but rather the process by which form takes its shape. Pioneering work by Gramazio Kohler Research at ETH Zürich such as the Endless Wall [46] and Flight Assembled Architecture [47, 40] explored the creative, aesthetic, and material implications of on-site robotic construction. In the Endless Wall, an articulated robotic arm on a mobile base tirelessly performed the Sisyphean task of continually reassembling a brick wall, while in Flight Assembled Architecture a team of autonomous quadcopters assembled a tall structure from polyurethane building modules. As a performance, the Endless Wall expressed an algorithmic aesthetics of infinity through its continual cycle of construction and deconstruction. The University of Stuttgart's Institute for Computational Design has pioneered construction processes for novel materials through a series of experimental pavilions including the Elytra Filament Pavilion, which was built from robotically wound filaments of carbon fiber [38, 30, 12]. In these projects robotic construction played a generative role in the design, with technique and material critically informing the creative form-finding process. The interdisciplinary practice MAEID explores the entanglement of technology and nature through creative experiments in robotic biofabrication such as Pahoehoe Beauty (Figure 1.8) [16, 17, 9], Magic Queen (Figure 1.9) [4], and Sylva (Figure 1.10) [5]. In Sylva, robotic gardeners tend an artificial ecosystem. The robots, using machine vision to interpret their environment, are programmed to seed, water, or trim the plants. While the plants rely on the robots for care, the robots – acting in response to their environment – only move in response to the plants. This performative installation asks how technology and nature can coexist in symbiosis, exploring themes of empathy, coexistence, non-human subjectivity, and distributed agency.

Figure 1.11: Dominic Jud, Ilmar Hurkxkens, & Marco Hutter, *Robotic Embankment*, 2020

1.3.2 Landscape Architecture

Landscape architects have used algorithms to rapidly model design alternatives for site systems such as the pathways for Fletcher Studio's South Park [27], the planting at Il Nature's Les Jardins d'Étretat [10], the paving patterns at Land Collective's Cummins Inc.'s Distribution Business Unit Headquarters [29], and the landforms at Snøhetta's MAX IV Laboratory Landscape [36]. For Cummins, Land Collective used an algorithmic approach to rapidly generate alternative paving patterns from fading gradients to the final design of alternating stripes of light and dark. This enabled the design team to quickly iterate through design concepts, testing many different ideas [7]. Computational design can also be used to build high performance landscapes. Snøhetta designed the MAX IV Laboratory Landscape for functional performance, algorithmically modeling and autonomously constructing landforms that would dampen ground vibrations around the particle accelerator and collect stormwater. GPS-controlled bulldozers were used to precisely construct these complex landforms. For Avena+ Test Bed, Benedikt Groß designed algorithmic patterns of crops, herbs, and flowers for wildlife friendly farming that were planted using a tractor with GPS-guided machine control (Figure 1.1) [43]. Using a Voronoi diagram, an eleven and half hectare field was partitioned into cells planted with oats and bordered with herbs and wildflowers to attract pollinators and predators for pests.

Landscape Processes While most buildings are essentially static once built, landscape is an inherently dynamic medium with plants growing and decaying, water flowing, and topography inexorably changing. Given the dynamic nature of their media, landscape architects have become interested in designing the physical and ecological processes that shape landscapes. Experiments in designing landscape processes include Field Operations' ecological succession plan for Freshkills Park [53] and Georges Descombes and Atelier Descombes Rampini's Renaturation of the River Aire [19]. In their draft masterplan for Freshkills Park – a massive, decommissioned landfill on Staten Island in New York that has been converted into a public park – Field Operations proposed establishing diverse habitats including wetlands, grasslands, and woodlands over the course of thirty years. The draft masterplan embraced the concept of ecological succession – the transition from one community, from one set of species and habitats, to another. For the grasslands, for example, fully established native prairies in wet soils would be allowed to transition to woodlands, while prairies in drier soils would be maintained with a disturbance regime. Every three years the prairies on drier land would be mown to cut down tree saplings that would outcompete the grasses. The disturbance regime would return above-ground biomass to the soil, helping to regenerate the degraded soils on the site. Tasked with renaturalizing the channelized River Aire, Georges Descombes and Atelier Descombes Rampini let the river shape itself rather than trying to restore it to a historical baseline or designing a naturalistic, but artificial, engineered morphology for the river. The designers had a new riverbed excavated and laid out a diamond grid of channels and islands. The new river found a shifting path through the grid, weaving its course through the islands, eroding and depositing its sediment in new formations. Within the first year, a braided river, superimposed over a dissolving grid, had formed.

Computational Processes The processes that shape landscapes can be modeled or simulated with algorithms from the spatial sciences. If used creatively, geospatial models can be generative systems for designing landscapes. With algorithmic form-finding based on geospatial modeling and simulation, landscape architects can computationally design physical and ecological processes, rather than merely designing static forms. Experiments in the computational design of dynamic, geomorphologically active landscapes include Alexander Robinson's Rapid Landscape Prototyping Machine [25], Cantrell and Holzman's Sedimachine [41], Petrasova et al.'s Tangible Landscape [22], and the Swiss Federal Institute of Technology in Zürich's Robotic Embankment (Figure 1.11) and Robotic Landscape Studios [14, 8]. The Robotic Landscape Studios explored how computational design, geospatial simulation, and robotic fabrication can be used to adaptively design interventions for geomorphologically active landscapes shaped by landslides or fluvial processes. These studios explored how autonomous earthmovers could continually build new landforms that would redirect flows of water, sediment, and debris to protect nearby communities. Algorithmically designed landforms, meant to redirect and deposit debris, were evaluated using numerical simulations of debris flow. The landforms were designed for robotic construction by autonomous earthmovers using on-site material, balancing cut-and-fill. These studios envision the iterative construction of a landscape in response to ongoing geomorphic processes. In this landscape algorithmic processes – of design, modeling, simulation, and construction – interact with geomorphic processes. With computational design, landscape architects can design dynamic landscapes, conceived of and modeled as ongoing performances, as continual interactions between form and process. This book is intended to serve as a guide for designers who wish to work with the processes that shape landscapes, to artistically engage with ecology and geomorphology. It explores how algorithms can be used creatively to model, analyze, simulate, and design landscapes.

References

[1] [Software] Python Software Foundation, *Python* version 3.11.4, 2023. lic: Python Software Foundation License. url: https://www.python.org/, vcs: https://github.com/python.

[2] "algorism, n". In: *Oxford English Dictionary Online*. Oxford University Press, 2022. url: https://doi.org/10.1093/OED/8490909016.

[3] "algorithm, n". In: *Oxford English Dictionary Online*. Oxford University Press, 2022. url: https://doi.org/10.1093/OED/1019775631.

[4] MAEID. *Magic Queen*. 2021. url: https://maeid.com/works/magic-queen (visited on 08/18/2023).

[5] MAEID. *Sylva*. 2021. url: https://maeid.com/works/slyva (visited on 08/18/2023).

[6] Jared Tarbell. *Infinite Regeneration*. 2021. url: https://www.infinite.center/2021/01/27/infinite-regeneration/ (visited on 03/18/2022).

[7] Jessica Canfield. "The Scripted Surface: Parametric Modeling Aids the Design for a Complex Paving Pattern at a Corporate Campus". In: *Landscape Architecture Magazine* (Nov. 2020), pages 36–44. url: https://landscapearchitecturemagazine.org/2020/11/10/the-scripted-surface/.

[8] Ilmar Hurkxkens. "Robotic Landscapes: Topological Approaches to Terrain, Design, and Fabrication". PhD thesis. Zurich: ETH Zurich, 2020. doi: 10.3929/ethz-b-000451100.

[9] Daniela Mitterberger and Tiziano Derme. "Digital soil: Robotically 3D-printed granular biocomposites". In: *International Journal of Architectural Computing* 18.2 (2020), pages 194–211. doi: 10.1177/1478077120924996.

[10] Seth Putnam. "The Carved Garden: Creative Approaches to Maintenance Shape the Modern Topiary at Les Jardins d'Étretat". In: *Landscape Architecture Magazine* (Dec. 2020), pages 24–25.

[11] Martin Alvarez, Hans Jakob Wagner, Abel Groenewelt, Oliver Krieg, Ondrej Kyjanek, Lotte Aldinger, Simon Bechert, Daniel Sonntag, Achim Menges, and Jan Knippers. "The Buga Wood Pavilion – Integrative Interdisciplinary Advancements of Digital Timber Architecture". In: *ACADIA – Ubiquity and Autonomy: Proceedings of the 39th Annual Conference of the Association for Computer Aided Design in Architecture* (2019), pages 490–499. doi: 10.52842/conf.acadia.2019.490.

[12] Niccolò Dambrosio, Christoph Zechmeister, Serban Bodea, Valentin Koslowski, Marta Gil-Pérez, Bas Rongen, Jan Knippers, and Achim Menges. "Buga Fibre Pavilion: Towards an architectural application of novel fiber composite building systems". In: *ACADIA 2019 – Ubiquity and Autonomy: Proceedings of the 39th Annual Conference of the Association for Computer Aided Design in Architecture* (2019), pages 140–149. url: http://papers.cumincad.org/cgi-bin/works/paper/acadia19_140.

[13] Nadja Gaudillière, Romain Duballet, Charles Bouyssou, Alban Mallet, Philippe Roux, Mahriz Zakeri, and Justin Dirrenberger. "Large-Scale Additive Manufacturing of Ultra-High-Performance Concrete of Integrated Formwork for Truss-Shaped Pillars". In: *Robotic Fabrication in Architecture, Art and Design 2018*. Edited by Jan Willmann, Philippe Block, Marco Hutter, Kendra Byrne, and Tim Schork. Cham: Springer International Publishing, 2019, pages 459–472. isbn: 978-3-319-92294-2.

[14] Ilmar Hurkxkens and Mathias Bernhard. "Computational Terrain Modeling with Distance Functions for Large Scale Landscape Design". In: *Journal of Digital Landscape Architecture* 2.4 (2019), pages 222–230. doi: 10.14627/537663024.

[15] Rafael Lozano-Hemmer. *Babbage Lovelace*. 2019. url: https://www.lozano-hemmer.com/babbage_lovelace.php (visited on 03/18/2022).

[16] MAEID. *Pahoehoe Beauty*. 2019. url: https://maeid.com/works/pahoehoe-beauty (visited on 08/18/2023).

[17] Daniela Mitterberger and Tiziano Derme. "Soil 3D Printing". In: *ACADIA – Ubiquity and Autonomy: Proceedings of the 39th Annual Conference of the Association for Computer Aided Design in Architecture* (2019), pages 586–595. doi: https://doi.org/10.52842/conf.acadia.2019.586.

[18] Andreas Thoma, Arash Adel, Matthias Helmreich, Thomas Wehrle, Fabio Gramazio, and Matthias Kohler. "Robotic Fabrication of Bespoke Timber Frame Modules". In: *Robotic Fabrication in Architecture, Art and Design 2018*. Edited by Jan Willmann, Philippe Block, Marco Hutter, Kendra Byrne, and Tim Schork. Cham: Springer International Publishing, 2019, pages 447–458. isbn: 978-3-319-92294-2.

[19] Georges Descombes. *Aire: The River and Its Double*. Park Books, 2018. isbn: 9783038600817.

[20] Mark Foster Gage, Peter Eisenman, and Robert A.M. Stern. *Mark Foster Gage: Projects and Provocations*. Rizzoli, 2018. isbn: 9780847862092.

[21] Joris Laarman Lab. *MX3D Bridge*. 2018. url: https://www.jorislaarman.com/work/mx3d-bridge (visited on 01/12/2022).

[22] Anna Petrasova, Brendan Harmon, Vaclav Petras, Payam Tabrizian, and Helena Mitasova. *Tangible Modeling with Open Source GIS*. 2nd edition. Springer International Publishing, 2018. isbn: 978-3-319-89302-0. doi: 10.1007/978-3-319-89303-7.

[23] Davide Quayola. *Remains: Vallée de Joux*. 2018. url: https://quayola.com/remains-vallee-de-joux (visited on 09/10/2023).

[24] Ronald Rael and Virginia San Fratello. *Printing Architecture: Innovative Recipes for 3D Printing*. Princeton Architectural Press, 2018. isbn: 9781616897475.

[25] Alexander Robinson. *The Spoils of Dust: Reinventing the Lake that Made Los Angeles*. ORO Editions, 2018. isbn: 9781940743486.

[26] Philipp Eversmann, Fabio Gramazio, and Matthias Kohler. “Robotic prefabrication of timber structures: towards automated large-scale spatial assembly”. In: *Construction Robotics* 1.1-4 (2017), pages 49–60. issn: 2509-811X. doi: 10.1007/s41693-017-0006-2.

[27] Fletcher Studio. *South Park*. 2017. url: https://www.fletcher.studio/southpark (visited on 07/28/2021).

[28] Markus Giftthaler, Timothy Sandy, Kathrin Dörfler, Ian Brooks, Mark Buckingham, Gonzalo Rey, Matthias Kohler, Fabio Gramazio, and Jonas Buchli. “Mobile robotic fabrication at 1: 1 scale: the in situ fabricator”. In: *Construction Robotics* 1.1-4 (2017), pages 3–14. doi: 10.1007/s41693-017-0003-5.

[29] Land Collective. *Cummins, Inc. Distribution (DBU) Headquarters*. 2017. url: https://land-collective.com/projects/cummins-distribution-headquarters (visited on 07/28/2021).

[30] Marshall Prado, Moritz Dörstelmann, Achim Menges, James Solly, and Jan Knippers. “Elytra filament pavilion: Robotic filament winding for structural composite building systems”. In: *Fabricate 2017*. UCL Press, 2017, pages 224–231. isbn: 9781787350014. doi: 10.2307/j.ctt1n7qkg7.35.

[31] US Army Corps of Engineers. *Automated Construction of Expeditionary Structures (ACES)*. 2017. url: https://www.erdc.usace.army.mil/Media/Fact-Sheets/Fact-Sheet-Article-View/Article/1290247/automated-construction-of-expeditionary-structures-aces/ (visited on 01/12/2022).

[32] Sean Ahlquist. “Sensory material architectures: Concepts and methodologies for spatial tectonics and tactile responsivity in knitted textile hybrid structures”. In: *International Journal of Architectural Computing* 14 (1 2016), pages 63–82. doi: 10.1177/1478077115625525. url: https://doi.org/10.1177/1478077115625525.

[33] Mark Burry. “Robots at the Sagrada Família Basilica: A Brief History of Robotised Stone-Cutting”. In: *Robotic Fabrication in Architecture, Art and Design 2016*. Edited by Dagmar Reinhardt, Rob Saunders, and Jane Burry. Cham: Springer International Publishing, 2016, pages 2–15. isbn: 978-3-319-26378-6. doi: 10.1007/978-3-319-26378-6_1.

[34] Ron Culver, Julia Koerner, and Joseph Sarafian. “Fabric Forms: The Robotic Positioning of Fabric Formwork”. In: *Robotic Fabrication in Architecture, Art and Design 2016*. Edited by Dagmar Reinhardt, Rob Saunders, and Jane Burry. Cham: Springer International Publishing, 2016, pages 106–121. isbn: 978-3-319-26378-6. doi: 10.1007/978-3-319-26378-6_8.

[35] Kathrin Dörfler, Timothy Sandy, Markus Giftthaler, Fabio Gramazio, Matthias Kohler, and Jonas Buchli. “Mobile Robotic Brickwork”. In: *Robotic Fabrication in Architecture, Art and Design 2016*. Edited by Dagmar Reinhardt, Rob Saunders, and Jane Burry. Cham: Springer International Publishing, 2016, pages 204–217. isbn: 978-3-319-26378-6. doi: 10.1007/978-3-319-26378-6_15.

[36] Snøhetta. *MAX IV Laboratory Landscape*. 2016. url: https://snohetta.com/projects/70-max-iv-laboratory-landscape (visited on 07/28/2021).

[37] Philip Yuan. “Parametric Regionalism”. In: *Architectural Design* 86.2 (2016), pages 92–99. doi: 10.1002/ad.2029. url: https://doi.org/10.1002/ad.2029.

[38] Achim Menges. *Material Synthesis: Fusing the Physical and the Computational*. Architectural Design. Wiley, 2015. isbn: 9781118878378.

[39] Davide Quayola. *Pleasant Places*. 2015. url: https://quayola.com/pleasant-places (visited on 09/10/2023).

[40] Frederico Augugliaro, Sergei Lupashin, Michael Hamer, Cason Male, Markus Hehn, Mark W Mueller, Jan Sebastian Willmann, Fabio Gramazio, Matthias Kohler, and Raffaello D’Andrea. “The Flight Assembled Architecture installation: Cooperative construction with flying ma-

chines". In: *IEEE Control Systems Magazine* 34.4 (2014), pages 46–64. doi: 10.1109/MCS.2014.2320359.

[41] Bradley Cantrell and Justine Holzman. "Synthetic Ecologies: protocols, simulation, and manipulation for indeterminate landscapes". In: *ACADIA 14: Proceedings of the 34th Annual Conference of the Association for Computer Aided Design in Architecture*. Los Angeles, 2014, pages 709–718. url: http://papers.cumincad.org/cgi-bin/works/paper/acadia14_709.

[42] Fabio Gramazio and Matthias Kohler. *Made by Robots: Challenging Architecture at a Larger Scale*. Architectural Design. Wiley, 2014. isbn: 9781118918951.

[43] Benedikt Groß. *Avena+ Test Bed: Agricultural Printing and Altered Landscapes*. 2013. url: https://benedikt-gross.de/projects/avena-test-bed-agricultural-printing-and-altered-landscapes (visited on 07/28/2021).

[44] Ryoji Ikeda. *data.path*. 2013. url: https://www.ryojiikeda.com/project/datamatics/#datapath (visited on 08/18/2023).

[45] Nick Dunn. *Digital Fabrication in Architecture*. Laurence King Publishing, 2012. isbn: 978-1-85669-891-7.

[46] Volker Helm, Selen Ercan, Fabio Gramazio, and Matthias Kohler. "In-situ robotic construction: Extending the digital fabrication chain in architecture". In: *ACADIA 2012 - Synthetic Digital Ecologies: Proceedings of the 32nd Annual Conference of the Association for Computer Aided Design in Architecture* 2012-October (2012), pages 169–176. url: http://papers.cumincad.org/cgi-bin/works/paper/acadia12_169.

[47] Jan Willmann, Federico Augugliaro, Thomas Cadalbert, Raffaello D'Andrea, Fabio Gramazio, and Matthias Kohler. "Aerial Robotic Construction Towards a New Field of Architectural Research". In: *International Journal of Architectural Computing* 10.3 (2012), pages 439–460. doi: 10.1260/1478-0771.10.3.439.

[48] Mario Carpo. *The Alphabet and the Algorithm*. Cambridge, MA: MIT Press, 2011, page 169. isbn: 978-0262515801.

[49] dECOi Architects. *One Main*. 2011. url: http://www.decoi-architects.org/2011/10/onemain/ (visited on 07/28/2021).

[50] Snøhetta. *Tverrfjellhytta, Norwegian Wild Reindeer Pavilion*. 2011. url: https://snohetta.com/projects/2-tverrfjellhytta-norwegian-wild-reindeer-pavilion (visited on 07/28/2021).

[51] Laura U. Marks. *Enfoldment and Infinity: An Islamic Genealogy of New Media Art*. Leonardo. Cambridge, MA: MIT Press, 2010. isbn: 9780262014212.

[52] Ryoji Ikeda. *test pattern*. 2008. url: https://www.ryojiikeda.com/project/testpattern (visited on 03/18/2022).

[53] Field Operations. *Fresh Kills Park: Draft Master Plan*. Technical report. New York: City of New York, 2006. url: https://www1.nyc.gov/assets/planning/download/pdf/plans/fkl/dmp.pdf.

[54] Mario Carpo. "Tempest in a Teapot". In: *Log* 6 (2005), pages 99–106. issn: 15474690. url: http://www.jstor.org/stable/41765067.

[55] Bernard Cache. *Earth Moves: The Furnishing of Territories*. Edited by Michael Speaks. Translated by Anne Boyman. Writing Architecture. MIT Press, 1995. isbn: 9780262531306.

[56] Ludwig Glaeser. *The Work of Frei Otto*. Museum of Modern Art, 1972. isbn: 9780870703331. url: https://www.moma.org/calendar/exhibitions/2662.

2 Visual Programming

2.1 Grasshopper

Grasshopper is a visual programming interface for the 3D modeling program Rhinoceros [1]. In visual programming, algorithms are composed as diagrams. With Grasshopper, algorithms are composed diagrammatically as networks of nodes connected by wires. Each node represents either data – such as text or geometry – or a computational process – such as triangulating a mesh through a set of points. Wires relay data between these nodes. These diagrams clearly show the relationships between the building blocks of an algorithm – between its input, processes, and output. By developing algorithms visually rather than textually, artists and designers can start programming intuitively – simply by dropping nodes onto a canvas and wiring them together – without having to worry about the syntax of a language. Visual programming lets artists and designers focus on computational thinking and logic rather than syntax. Visual programming is used widely throughout music, art, design, and the spatial sciences. Artists and musicians use visual programming languages such as vvvv [6], Max [8], and Pure Data [10], while spatial scientists use visual programming interfaces for geographic information systems and remote sensing [4].

There are many resources for learning the basics of Grasshopper. The *Grasshopper Primer* is an excellent guide for getting started that clearly describes the interface, programming process, data types, and basic operations [9]. *Essential Mathematics for Computational Design* describes the key mathematics concepts – vectors, matrices and transformations, and parametric curves and surfaces – needed for constructing algorithms in Grasshopper [7]. *Essential Algorithms and Data Structures for Computational Design* explains how to design algorithms in Grasshopper and how to work with data structures like lists and data trees [5]. There are a plethora of tutorials for Grasshopper including great introductory videos such as "Grasshopper Getting Started By David Rutten" and Rajaa Issa's "Jump Start Grasshopper."

2.1.1 Interface

To compose algorithms with Grasshopper, drag and drop nodes from the components tab onto the canvas (Figure 2.1). The components panels contain parameters that collect and store data and components which run processes on data. The core panels include `Params`, `Maths`, `Sets`, `Vector`, `Curve`, `Surface`, `Mesh`, `Intersect`, `Transform`, `Display`, and `Kangaroo 2`. Installing extensions can add additional panels with new parameters and components (appendix A.1). Only the most commonly used components are shown in the panels; to see the complete set of components grouped in each panel, click on the dark gray bar at the bottom of a panel to expand it and reveal the full list. Browsing through the panels is a good way to learn what components are available. Once you have learned the name of a component, however, it is often quicker and easier to search for it. Simply double click on the canvas to search for parameters or components.

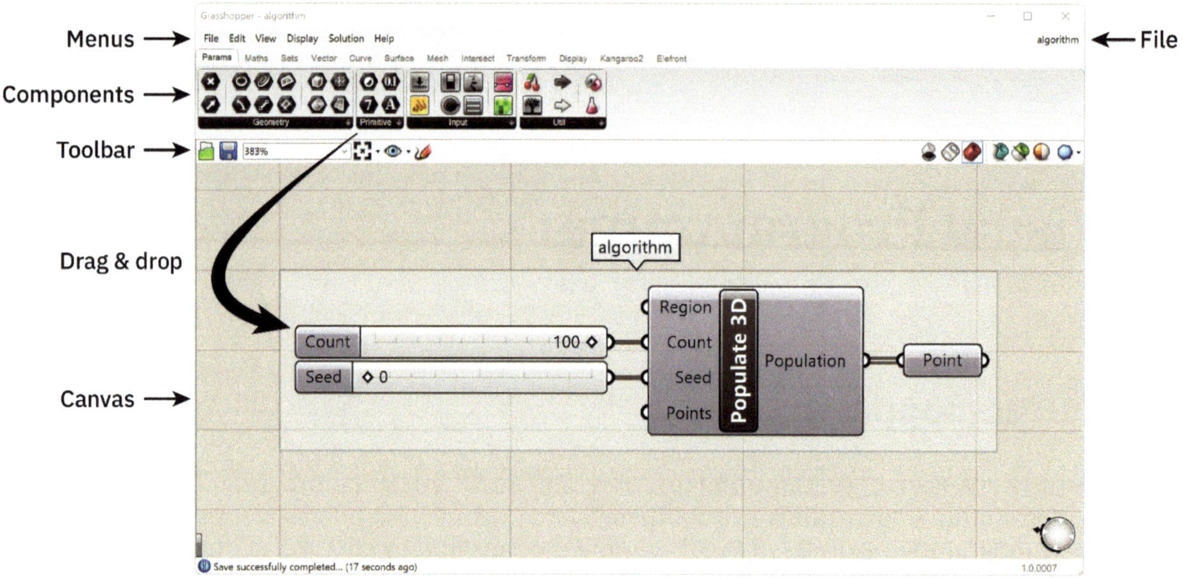

Figure 2.1: Grasshopper's interface

Nodes are color coded (Figure 2.2). Light gray nodes are working properly and will display a preview of their geometry in Rhino's viewports. Darker gray nodes are working properly, but their preview has been disabled. Nodes that are entirely grayed out have been disabled and will not process data. Right click on a node to turn its preview on or off or to enable or disable the node. Alternatively, use the middle mouse button to pull up a radial context menu with controls for previewing, enabling, grouping, zooming, clustering, baking, etc. When a node has a warning or an error it will be colored orange or red respectively with a balloon in the top right corner reporting the warning or error message. Unless they have default values, nodes will appear orange when first placed on the canvas because they need to be assigned input data. Without any input, components cannot run and will generate warnings. With the wrong type of input data, components will not be able to process the data and will generate errors.

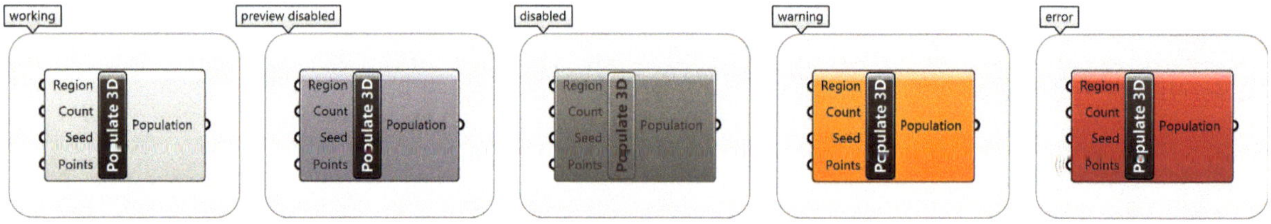

Figure 2.2: Nodes

2.1.2 Algorithms

An algorithm is a set of clearly defined, sequential rules for completing a task. The building blocks of an algorithm are inputs, processes, and outputs [5]. To compose an algorithm with Grasshopper, plug input parameters that collect and store data into components that process the data, and then plug the results into output parameters (Figure 2.3). Constructing a complex algorithm may require plugging many different inputs into a chain of components that run a sequence of processes. Algorithms in Grasshopper run from left to right, with input on the left of each node and output on the right. Data, thus, flows between nodes from left to right [9].

Parameters are containers for data. Input parameters – such as number sliders, value lists, Boolean toggles, color gradients – are interactive so you can easily set and change the data they store. Other parameters simply store different types of data broadly categorized as primitives – including text, integers, floating point numbers, Boolean truth values, and colors – and geometry – including points, curves, meshes, surfaces, and boundary representations. Components run processes on data. Grasshopper's built-in components include mathematical operations, list manipulation, geometric operations, intersections, transformations, and physics.

Grasshopper automatically compiles your program as soon as it changes, rendering a preview of the resulting geometry in real time. As soon as a component has input data, it runs and a preview of the resulting geometry is rendered in Rhino's viewports. After any change to a program – such as moving a number slider – all downstream processes will be solved again. Automatic compilation can be turned off by disabling the Grasshopper's solver in the `Solution` menu. The Grasshopper geometry rendered in Rhino is just a preview; it needs to be baked to become actual geometry in Rhino. To bake Grasshopper geometry, right click on an output parameter to pull up a context menu and select `Bake`.

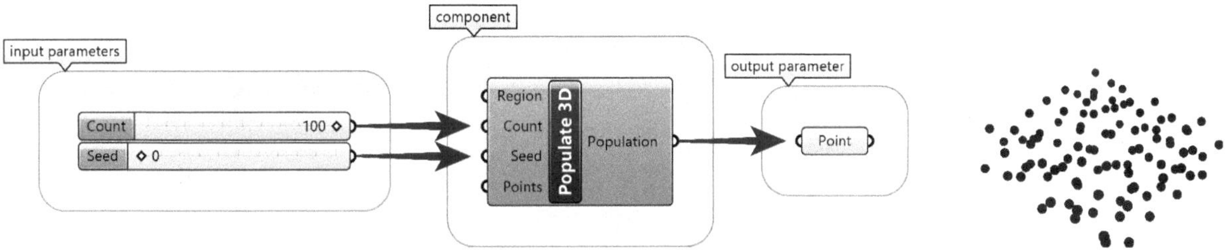

Figure 2.3: Composing an algorithm in Grasshopper

The visual complexity of large programs can undermine the intuitive nature of visual programming. Visual programs can quickly become messy tangles of wires. A messy program is hard to read and thus hard to use, develop, or troubleshoot. Strategies for dealing with the visual complexity of definitions in Grasshopper include using a systematic layout, staying organized, utilizing modularization, hiding wires, passing parameters, and writing good documentation [2]. Large programs can be divided into modules – into smaller, more manageable parts using groups, clusters, or hops. Groups are collections of nodes with outlines drawn around them that can easily be moved around together. To create a group, select the desired nodes and click `Ctrl+G`. The wires connecting components can be hidden. To hide a wire coming into a component, right click on the component's input parameter and set the wire display to hidden. Hidden wires are useful for passing parameters around the canvas. For example, when reusing an input parameter, rather than drawing a second wire a long way across the canvas, store it in another parameter, hide the wire, and move the copy across the canvas. This strategy is especially useful for passing parameters between groups (Figure 2.4). Scribble parameters can be used to jot down notes, write documentation, or add labels directly on the canvas. Relays can be used to control the path of wires. Simply double click on a wire to add a relay. Extensions like the Snapping Gecko addon make it easier to neatly align nodes [3]. These stylistic strategies are used throughout this book to keep the visual programs neat, organized, and legible.

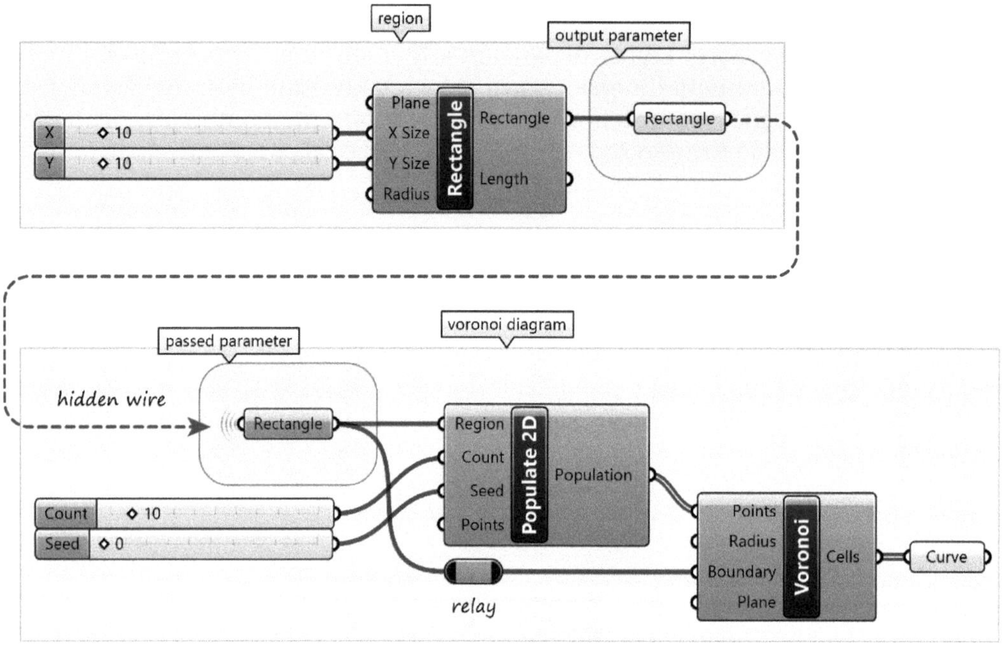

Figure 2.4: Modularization

2.2 Geometry

Basic types of geometry in Grasshopper include points, curves, meshes, surfaces, boundary representations (BReps), subdivision surfaces (SubD), and vectors. Grasshopper in Rhino 8 also supports point clouds. Extensions can enable other types of geometry such as voxels, i.e. volumetric pixels. Use the Grasshopper definition `fundamentals.gh` as a guide for this chapter.

Code: `fundamentals.gh`

2.2.1 Points

Points are positions in space. In a 3-dimensional Euclidean space, points are represented as sets of x-, y-, and z-coordinates. Points are the geometric foundation for the computational design of landscapes as they are used to construct other geometries such as curves, meshes, and surfaces. While terrain, for example, may be modeled as a raster, mesh, or surface, these geometries are derived from a set of points – a point cloud. Spatial data from global navigation satellite systems, surveying, lidar, and photogrammetry are collected as points.

In Grasshopper use the `Construct Point` component to construct a single point from a set of coordinates. Plug number sliders into the input x-, y-, and z-coordinates to interactively set the position of the point. Points modeled in Rhino can also be referenced in Grasshopper. To do this, place a `Point` parameter, right click on it and select `set one point` from the context menu, and then select an existing point in Rhino's viewports. To save a referenced point as part of the Grasshopper definition, right click on the `Point` parameter and select `internalize data` from the context menu. `Point` parameters with internalized data can be transformed with Grasshopper's gumball. The gumball can be enabled in the Display menu.

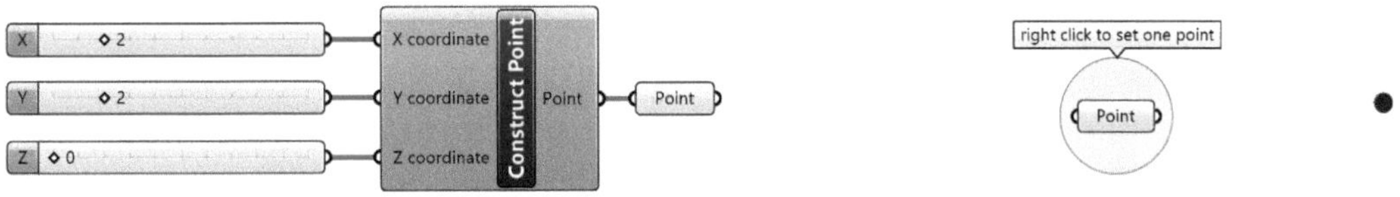

A random population of points can be generated using the `Populate 2D`, `Populate 3D`, or `Populate Geometry` components. Place a `Populate 3D` component and plug number sliders into the count and seed input parameters. The count parameter determines the number of random points that will be generated, while the seed parameter initializes the pseudo-random number generator that sets the coordinates of the points. Changing the seed will generate a new set of points with different positions in space.

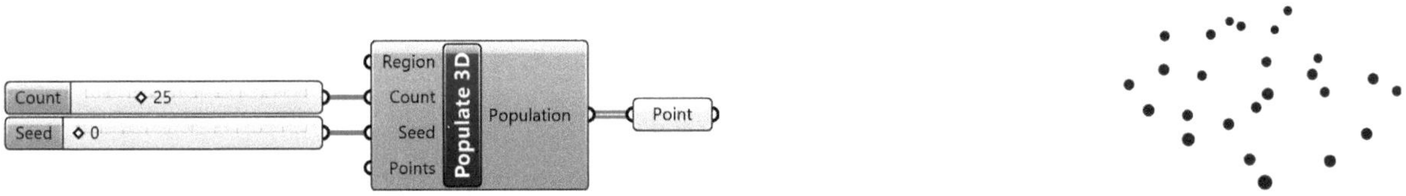

2.2.2 Curves

In Rhino, curves are modeled as non-uniform, rational-basis splines that are shaped by control points. A curve is divided into a sequence of knots. Control points pull the knots, shaping the curve. The degree or order of the curve is the number of control points that influences each knot. Curves can be interpolated through a set of points that become knots with the `Interpolate` component or constructed from a set of control points with the `Nurbs Curve` component. To draw a curve through a random set of points, start by placing a `Populate 3D` component and plug in number sliders for the count and seed input parameters. Then plug the output population of points into an `Interpolate` component as the input vertices.

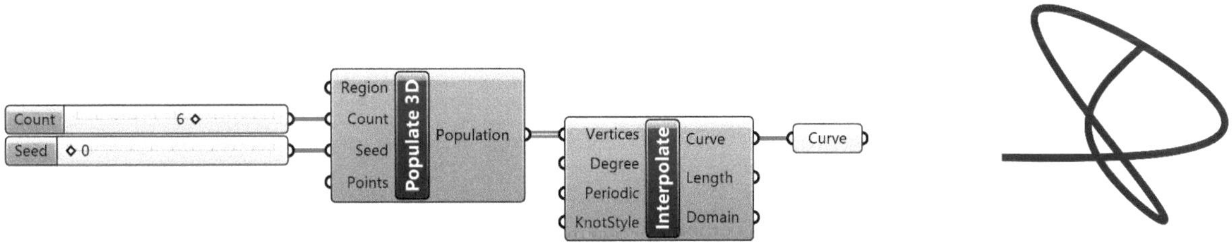

2.2.3 Meshes

Polygonal meshes are networks of discrete, connected cells composed of vertices, edges, and faces. Meshes composed of triangular cells – referred to as triangulated irregular networks (TINs) in the spatial sciences – are often used to model terrain. Delaunay triangulation is a method for generating meshes with good aesthetics by avoiding narrow cells. In Delaunay triangulation, a set of points is connected as a mesh with maximized minimum interior angles in order to avoid thin triangular cells. Use the `Delaunay Mesh` component to generate a triangulated mesh in Grasshopper. First generate a random distribution of points in 3-dimensional space with the `Populate 3D` component. Then plug the output population of points into the `Delaunay Mesh` component as the input points parameter.

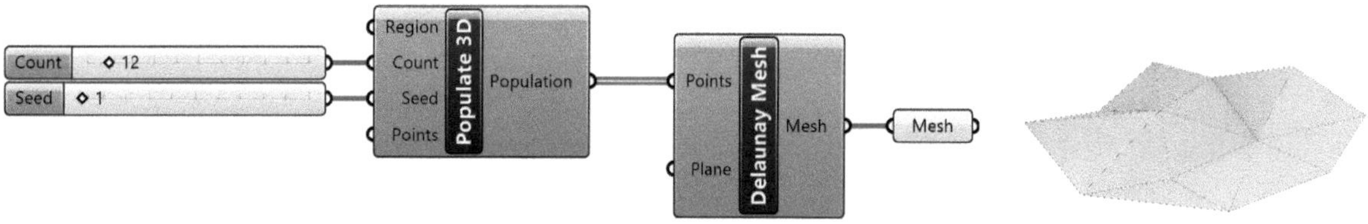

2.2.4 Surfaces

A NURBS surface is interpolated through a set of knots that are pulled by control points. Knots are connected by knot spans that can be rendered as isocurves. Components for generating surfaces include `Boundary Surface`, `Patch`, `Ruled Surface`, `Loft`, `Network Surface`, and `Surface from Points`.

The `Patch` component generates a surface from a set of points and curves. A planar surface with a given number of spans is fitted through these points and curves and then deformed to match. The reference geometry acts as control points pulling on the knots to shape the surface. How well the surface matches the reference geometry depends on its flexibility and the number of knot spans specified. Use the `Patch` component to create a surface from a random population of 3D points. First generate the points with a `Populate 3D` component. Then plug the output population of points into a `Patch` component as the input `Points` parameter. Set the number of spans with a number slider.

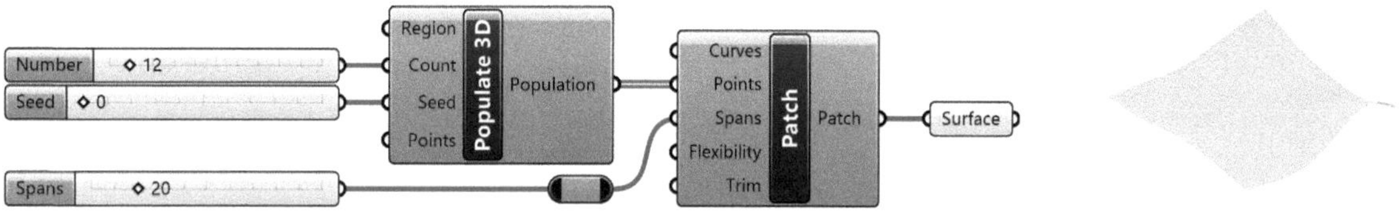

2.2.5 Boundary Representations

In Rhino, complex geometry such as solid volumes can be precisely modeled as Boundary Representations (BReps). This data structure is defined topologically as faces, edges, and vertices, and geometrically as surfaces, curves, and points. A simple way to create a solid volume from a surface is to extrude it. Start by modeling a surface (section 2.2.4). Plug the surface into an `Extrude` component. To define the direction of extrusion, plug a `Unit Z` component in as the input `Direction` parameter. To set the amplitude of the vector, plug a number slider into the `Unit Z` component. Finally plug the resulting extrusion into a BRep parameter to convert it.

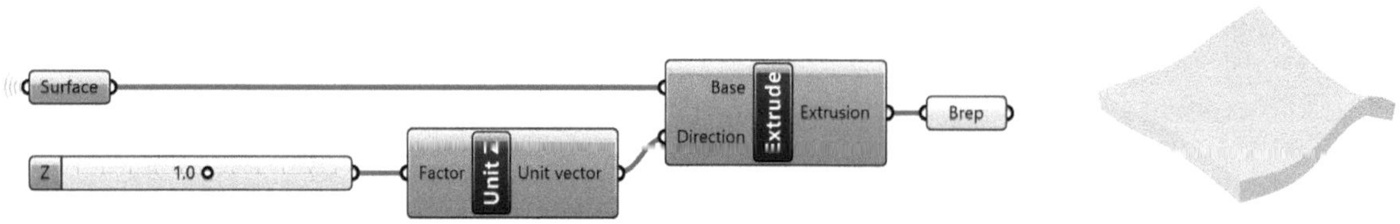

2.2.6 Vectors

A vector is a direction and magnitude of motion. The unit vectors $\vec{i}$, $\vec{j}$, and $\vec{k}$ represent a single unit of motion along the x-, y-, and z-axes of 3-dimensional Cartesian space. In Grasshopper these vectors can be created with the `Unit X`, `Unit Y`, and `Unit Z` components and visualized with `Vector Display` components.

Vectors can be constructed by defining their magnitude of motion in the *x*, *y*, and *z* dimensions with the `Vector XYZ` component.

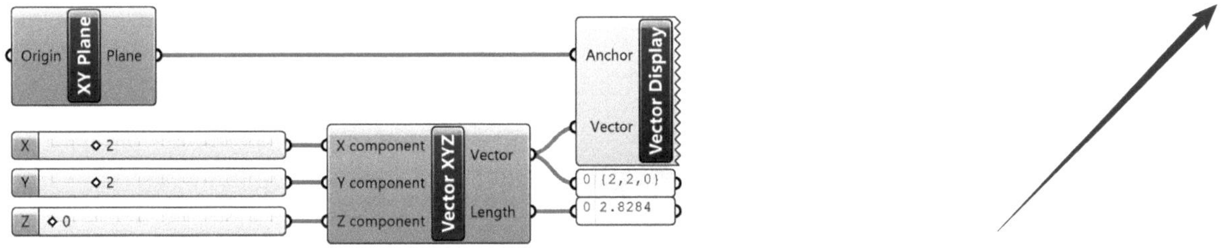

Vectors can also be defined as the motion between a starting point and an ending point with the `Vector 2Pt` component.

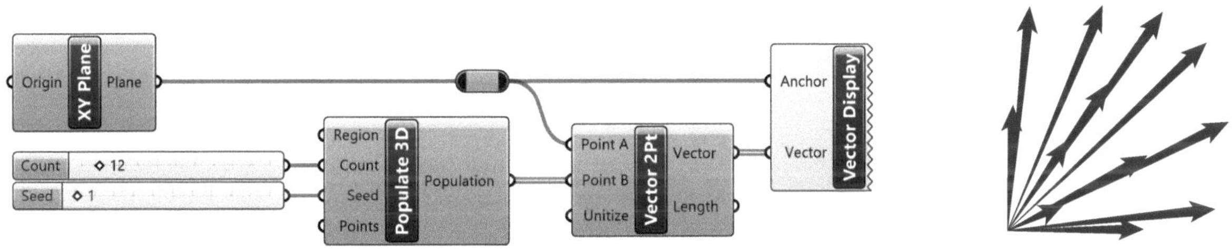

2.3 Transformations

Basic transformations in Grasshopper include translation, rotation, and scaling.

2.3.1 Translation

Translation is moving geometry along a vector. In Grasshopper, geometry can be translated with the `Move` component. This component requires an input geometry to move and an input vector to move along. Translate a box along the unit-x vector. First generate a box with the `Center Box` component. Specify its x-, y-, and z-dimensions with number sliders. Then plug the box into a `Move` component as the input geometry. Define the translation vector by setting a `Unit X` component as the input `motion`. Define the amplitude of the translation vector by plugging a number slider into the `Unit X` component as the input `factor`.

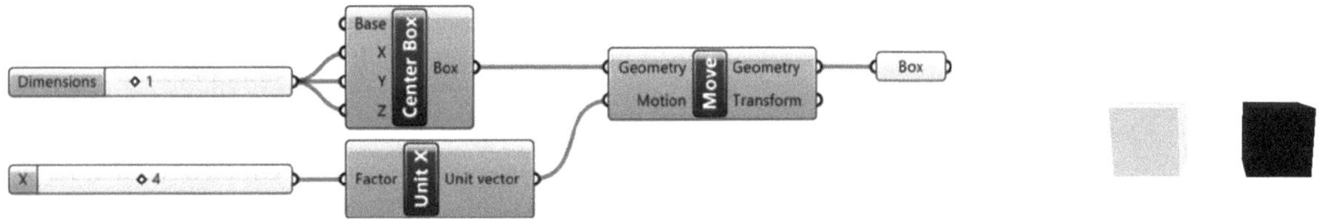

2.3.2 Rotation

In Grasshopper, geometry can be rotated with the `Rotate` component. This component requires an input geometry, rotation angle in radians, and rotation plane. Rotate a box around its center on the XY-plane. In this example the box is specified by a `Box` parameter with a hidden wire to the translated box in the previous example. First plug a box into a `Rotate` component as the input geometry. To specify the input rotation angle in radians, plug a number slider with two decimal places into a `Pi` component from the math tab and plug this into the `Rotate` component as the input angle. This component uses the XY world plane as its default rotation plane. To specify the center of the box as the origin of the rotation plane, plug the box into an `Area` component to find its centroid. Then plug the output centroid into the `Rotate` component as the input plane.

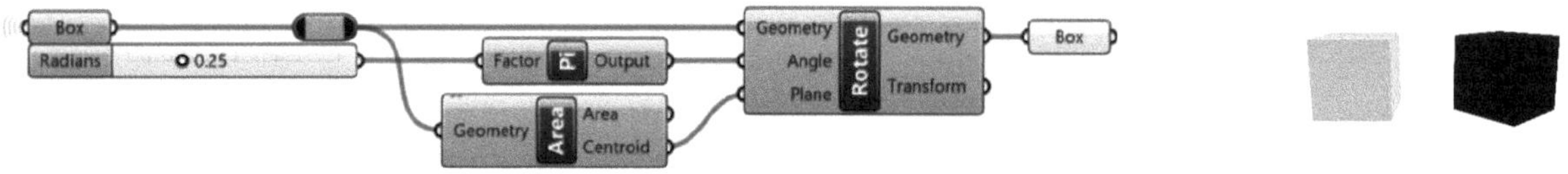

To set the rotation angle in degrees instead of radians, right click on the angle input and check the `degrees` option.

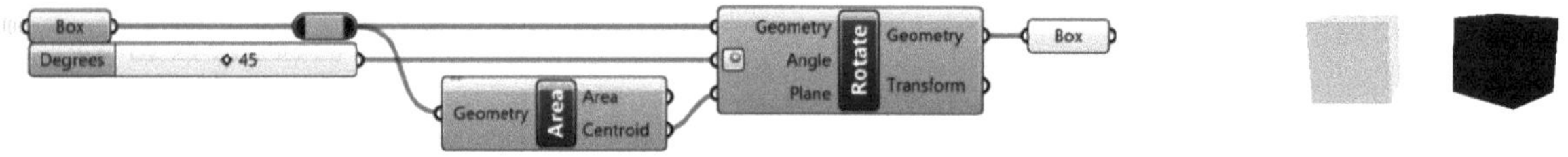

2.3.3 Scaling

Geometry can be scaled uniformly in Grasshopper with the `Scale` component. This component requires an input geometry, center of scaling, and scale factor. To scale a box from its center, plug the box into a `Scale` component as the input geometry. To set the scale factor, plug the box into an `Area` component to find its centroid and then plug the centroid into the `Scale` component as the input center. Finally plug a number slider in as the input factor.

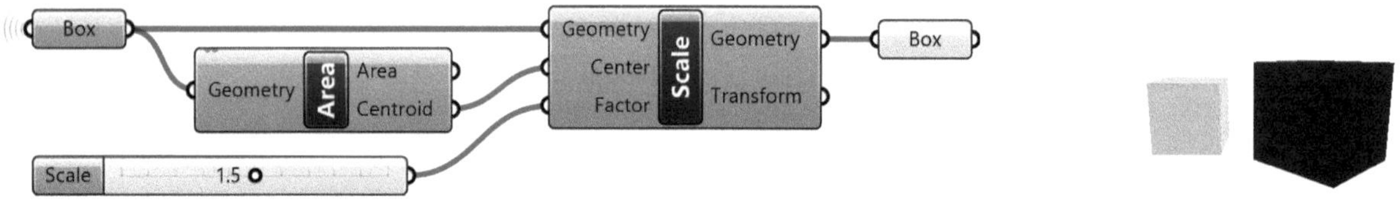

2.4 Data Structures

Key data types and data structures in Grasshopper include strings, numbers, Booleans, lists, and data trees.

2.4.1 Strings

Strings represent text. Strings are sequences of characters such as letters, numbers, symbols, and white space. In Grasshopper strings can be created easily using `Panels`. Simply place a panel and double click on it to enter text.

2.4.2 Numbers

Numbers can be represented as either integers or floating point numbers. Integers are whole numbers such as `1`, while floating point numbers are decimals such as `1.0`. In Grasshopper, numbers can be created easily with number sliders.

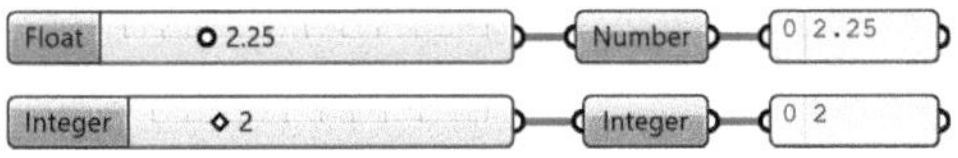

2.4.3 Booleans

Booleans are truth values that are either true or false. In Grasshopper, they can be set with `Boolean Toggle` parameters. Simply double click on the dark grey button to toggle the value between `True` and `False`. To learn more about Booleans and logical operations read chapter 1.3.5 of the Grasshopper Primer [9].

2.4.4 Lists

A list is an ordered sequence of data. Each item in a list has an index – a number that records its position in the list. Lists can be made manually in Grasshopper by placing a panel, right clicking to set multiline data, and entering each list item on a new line. List can also be made procedurally using components such as `Series`. To create a list of numbers, simply place a `Series` component on the canvas and set the count with a number slider. Plug the result into a panel to see the list. The first column contains indices, while the second column contains the items in the list.

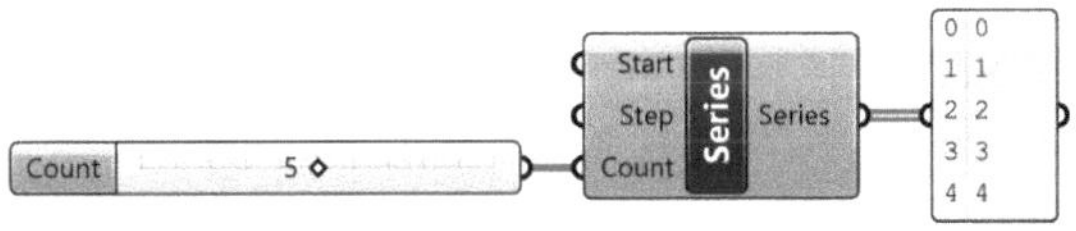

2.4.5 Data Trees

Data trees are nested lists, i.e. lists of lists. This data structure is unique to Grasshopper. Data trees are composed of a hierarchy of branches. Branches may branch into other branches, but eventually they end with data – with an item. An item or branch's position within a data tree is called a path. These are shown in panels as grey headers. Create a data tree of numbers by plugging a `Series` component into another `Series` component. Set the first `Series` to begin at one instead of the default zero. Plug the resulting list into the second `Series` component to create a data tree. The first list will become the branches of the tree, while the second list will become the data at the end of the branches. Visualize the data tree with a panel and a `Param Viewer`; right click on the viewer to display the tree as a graph.

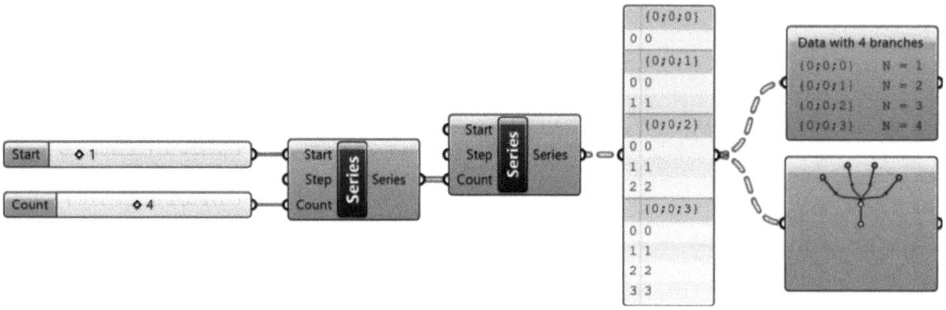

2.5 List Management

Working with lists and data trees is important part of Grasshopper. Methods for manipulating these data structures include flattening, grafting, culling, dispatching, and weaving.

2.5.1 Flatten

A data tree can be flattened into a list. Data in different branches will be merged into a single list. To flatten a data tree right click on a parameter or a component's input or output that the data tree passes through and set `Flatten`. Flattened data is represented by an icon with a downward facing arrow.

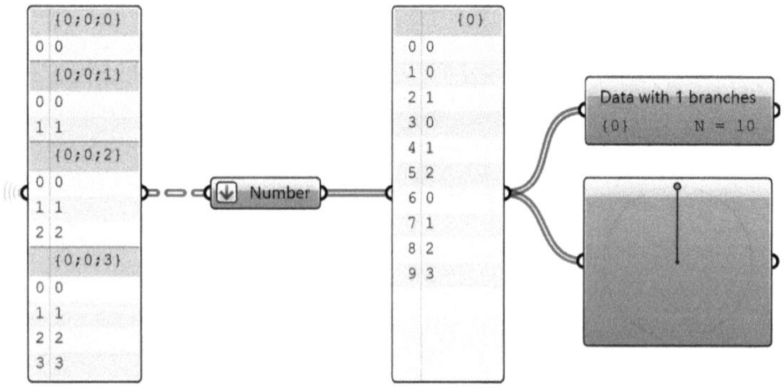

2.5.2 Graft

A list can be grafted into a data tree. Grafting a list converts each item into a branch. To graft a list, right click on a parameter, input, or output and set `Graft`. Grafted data is represented by an icon with an upward facing arrow.

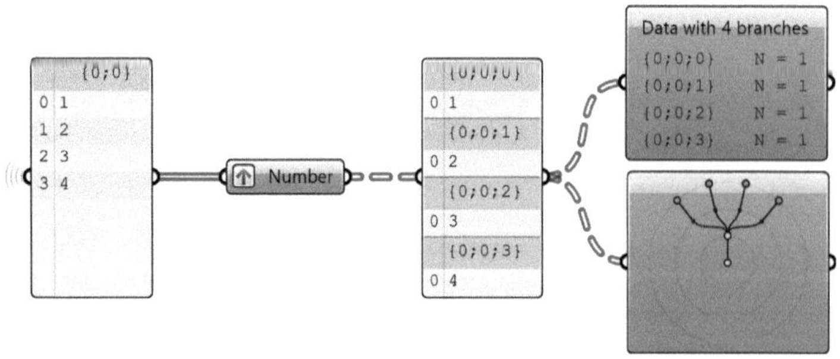

2.5.3 Culling Pattern

Items can be removed from lists with components such as `Cull Pattern`, `Cull Nth`, and `Random Reduce`. Use the `Cull Pattern` component to remove items from a list based on a pattern of Booleans. Start by creating a list of numbers with the `Series` component. Then plug the resulting list into a `Cull Pattern` component. Set the culling pattern using multiple `Boolean Toggles`. Remember to hold shift when plugging in multiple inputs.

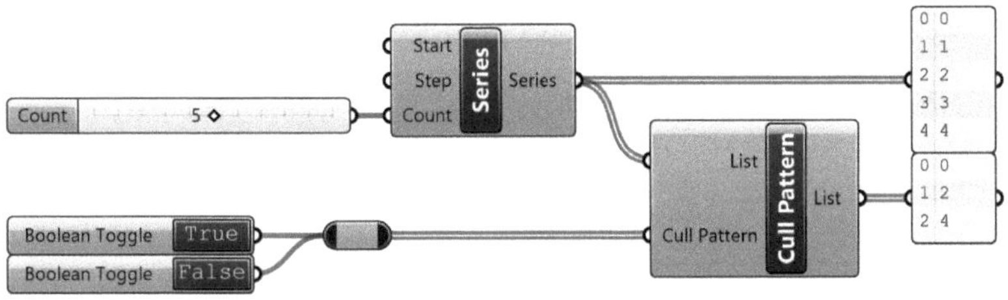

2.5.4 Dispatch

The `Dispatch` component splits a list into a set of lists based on a pattern of Booleans. Use `Dispatch` to create a list of even numbers and another list of odd numbers. Start by creating a list of numbers with the `Series` component. Then plug the resulting list into a `Dispatch` component and set the dispatch pattern using multiple `Boolean Toggles`. Create an alternating pattern by setting the first toggle to `True` and the second to `False`.

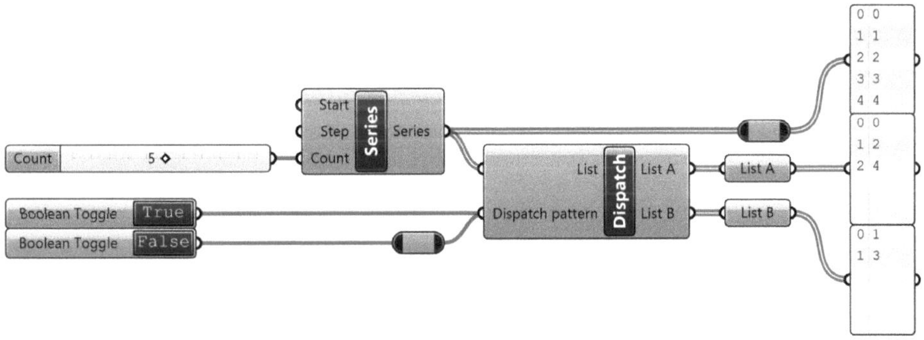

2.5.5 Weave

With the `Weave` component, separate streams of data can be woven together into a single stream based on pattern of Booleans. Create a list with a sequence of whole numbers where the even numbers are positive and the odd numbers are negative. Start by placing two integer parameters. Rename the first parameter to `List A` and plug in the list of even numbers generated by `Dispatch` in the previous example (section 2.5.4). Rename the second parameter to `List B` and plug in the list of odd numbers. Place a Boolean parameter and plug in both of the Boolean toggles from the previous example. Right click on each of the parameters and hide their wires. Make `List B` – the list of odd numbers – negative by plugging it into a `Negative` component. Then place a `Weave` component and plug the list of Booleans in as the input pattern, the list of positive, even numbers as the first stream, and the list of negative, odd numbers as the second stream.

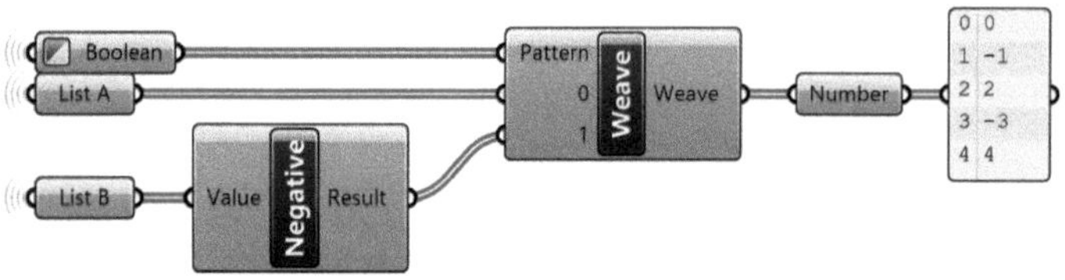

These methods for managing lists can be used to generate geometry. A wave that grows in amplitude, while decreasing in frequency can easily be created, for example, from a single sequence of numbers using methods like `Dispatch` and `Weave`. Start by generating a sequence of numbers with the `Range` component. Plugging this sequence of numbers into a `Construct Point` component would create a set of x-coordinates spaced at equal intervals along the x-axis. To gradually increase in the spacing of the points along the x-axis and thus decrease the frequency of the wave, compute the cumulative sum of the sequence with the `Mass Addition` component. This will add each value in the sequence to all of the preceding values. Plug the output of `Range` into a `Mass Addition` component and then plug its partial results into a `Construct Point` component as the x-coordinates. Meanwhile generate a sequence of z-coordinates that alternate between negative and positive values, while increasing in magnitude. First split the sequence of numbers generated by the `Range` component into two lists using `Dispatch` with the default, alternating pattern. Make the values in the second list negative with a `Negative` component. Then recombine both lists using `Weave` with the same default pattern and plug the resulting list into `Construct Point` as the z-coordinates. Finally interpolate a curve through this set of points with an `Interpolate` component.

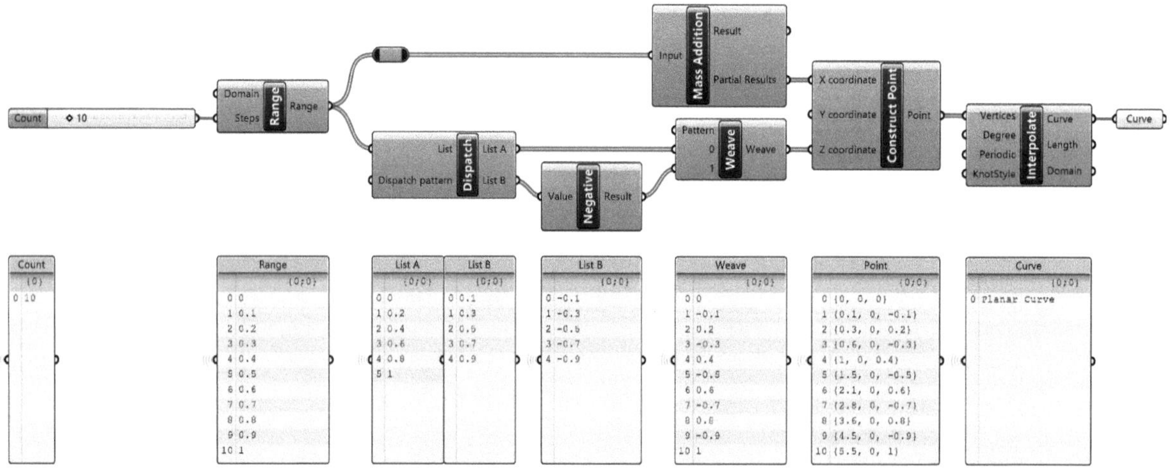

References

[1] [Software] Robert McNeel & Associates, *Rhinoceros* version 8, 2023. url: https://www.rhino3d.com/.

[2] Delft University of Technology. *Grasshopper "Good Practice"*. TOI-Pedia. 2021. url: http://wiki.bk.tudelft.nl/toi-pedia/Grasshopper (visited on 09/04/2022).

[3] [Software] weArk, *SnappingGecko* version 1.1.2, 2021. lic: CC-BY-SA 4.0. url: https://www.we-ark.fr/, vcs: https://www.food4rhino.com/en/app/snappinggecko.

[4] Zdena Dobesova. "Visual Programming for GIS Applications". In: *The Geographic Information Science & Technology Body of Knowledge*. Edited by John P. Wilson. University Consortium for Geographic Information Science, 2020. doi: 10.22224/gistbok/2020.1.7.

[5] Rajaa Issa. *Essential Algorithms and Data Structures for Computational Design*. Robert McNeel & Associates, 2020. url: https://www.rhino3d.com/download/rhino/6.0/essential-algorithms.

[6] [Software] vvvv group, *vvvv* version gamma, 2020. url: https://vvvv.org/, vcs: https://github.com/vvvv.

[7] Rajaa Issa. *Essential Mathematics for Computational Design*. Robert McNeel & Associates, 2019. url: https://developer.rhino3d.com/guides/general/essential-mathematics/.

[8] [Software] Cycling '74, *Max* version 8, 2018. url: https://cycling74.com/products/max.

[9] Mode Lab. *The Grasshopper Primer*. 3rd edition. Robert McNeel & Associates, 2015. url: http://grasshopperprimer.com.

[10] Miller Puckette. "Pure Data: another integrated computer music environment". In: *Proceedings of the second intercollege computer music concerts* (1996), pages 37–41.

3 Programming

3.1 Python

While artists and designers can intuitively build algorithms with visual programming, programming languages like Python may be needed to develop more complex algorithms. In this book computational thinking is taught first through visual programming and then with Python. Many of the computational design processes explored in this book will be presented first with visual programming diagrams and then with code.

Python is a free, open source programming language that is very powerful, yet easy to learn [6]. It is a high-level language – a programming language that focuses on logic and concepts, rather than on interactions with hardware. As a high-level language with simple syntax and dynamic semantics, it is easy to read and thus learn. It supports many programming paradigms including object-oriented programming for modularity and reuse. Python was designed to be easy to extend with modules and packages [14]. In addition to a large standard library of built-in modules, thousands of modules have been developed by the community. There are libraries such as NumPy [5, 18] and SciPy [12, 20] for numerical and scientific programming, Pandas [19, 25] for statistics, Matplotlib [13, 26] and Seaborn [15, 17] for data visualization, scikit-learn [11, 23] and TensorFlow [10] for machine learning, the Geospatial Data Abstraction Library [7], Shapely [8], and Geopandas [9] for spatial science, and Compas [2] for architecture.

In this book, Python scripts will be run either in an integrated development environment – a program for writing, running, and debugging code – or in Rhino using the ScriptEditor environment.

3.1.1 Conda

This book uses the `Miniconda` [1] distribution of Python because it is easy to install and get started. This distribution includes CPython, the standard implementation of Python and a package manager called Conda [3]. Conda can install Spyder [4] – an integrated development environment where you can write, run, and debug code and visualize the results – and a vast collection of packages for data science. See appendix A.2 for a guide to installing Miniconda, Spyder, and the packages needed for this book.

3.1.2 Hello World

Write your first Python script using the Spyder integrated development environment. Use Python's built-in `print` function to write a short message. Functions are reusable blocks of code that take input parameters and when run return output data. To call a function, write the name of the function followed immediately by a set of parentheses. Specify any input parameters inside of these parentheses. In this example the input parameter will be your message. Because your message will be text, write it inside of quotation marks. The quotation marks specify that it is a string, a datatype that stores characters. To run the script, simply hit the run button in Spyder and your message will be displayed in the console.

```python
print("Hello World")
```

```
Hello World
```

3.1.3 Hello Worlds

Use a `for` loop to print each item in a list. Use the script `hello-worlds.py` as a guide.

Code: `hello-worlds.py`

Lists In Python lists are ordered, changeable collections of data. They can contain data of different types. Lists are defined by ordered sequences of items separated by commas, all enclosed by square brackets and then assigned to a variable. Create a list of planets with the following statement:

```python
worlds = ["Earth", "Gallifrey", "Arrakis", "Trantor", "Tatooine"]
```

Loops A `for` loop iterates, repeatedly running a process until its conditions are met. The header of the loop defines the conditions, while the body defines the processes. Use a `for` loop to print a message about each planet in the list of planets. Define the header of the loop with the statement `for world in worlds:`. A colon marks the end of the header. In Python's syntax the body of a loop is indented to show that is a discrete unit of code. Define the body of the loop by indenting once and then writing a print statement such as: `print(world)`.

```python
worlds = ["Earth", "Gallifrey", "Arrakis", "Trantor", "Tatooine"]

for world in worlds:
    print(world)
```

Concatenation Concatenation is a simple way to format strings by joining them together. To concatenate two strings in Python, simply write a plus between the strings. Variables containing strings can also be concatenated. Try concatenating a greeting with each of your planets' names using the statement: `print("Hello " + world)`.

```python
worlds = ["Earth", "Gallifrey", "Arrakis", "Trantor", "Tatooine"]

for world in worlds:
    print("Hello " + world)
```

F-string Interpolation Use string formatting to write a more complex print statement. F-string – formatted string literal – interpolation is a more powerful, elegant way to format strings. With f-strings Python expressions can be embedded inside of strings. Type `f` before the opening quotation mark of a string to designate that it will be an f-string. Use curly brackets `{}` to include a Python expression inside of the f-string. Use the statement `print(f"Hello {world}")` to embed the variable `world` inside the string.

```python
worlds = ["Earth", "Gallifrey", "Arrakis", "Trantor", "Tatooine"]

for world in worlds:
    print(f"Hello {world}")
```

Comments Use comments to clearly describe your code. To write a comment simply start a new line with a # hash symbol followed by your text.

```
# set list of worlds
worlds = ["Earth", "Gallifrey", "Arrakis", "Trantor", "Tatooine"]

# print list of worlds
for world in worlds:
    print(f"Hello {world}")
```

```
Hello Earth
Hello Gallifrey
Hello Arrakis
Hello Trantor
Hello Tatooine
```

3.1.4 Hello Point

Python can also be used to generate and visualize geometry. Start by plotting the simplest of geometry – a point – a set of x-, y-, and z-coordinates in Cartesian space. This example uses the Matplotlib visualization library [26] to render a single point as a 3-dimensional scatterplot. To render a point, first import the library, then assign variables, create an empty figure, and plot the coordinates on the figure.

Code: `hello-point.py`

Import Start a new script in Spyder. First import the `matplotlib` library with the statement `import matplotlib`. Libraries and modules need to be imported before any other code is run.

Variables Assign each coordinate of the point to a variable with the statements `x = 0`, `y = 0`, and `z = 0`. Write each statement on a new line. Just like in mathematics, variables are symbols representing objects. In programming, a variable has a name and stores data. Variables make it easy to store and retrieve things in programming. In Python, variables are assigned with an = equals symbol.

Plotting Create a figure with the statement `ax = matplotlib.pyplot.subplot(projection = '3d')`. The function `matplotlib.pyplot.subplot` creates the figure as a Python object. Setting `subplot()`'s `projection` parameter to `'3d'` makes the figure 3-dimensional. The object is then assigned to the variable `ax`. Objects like `ax` can store data as attributes and call functions as methods. Plot the coordinates on the figure as a scatterplot with the statement `plot = ax.scatter(x, y, z)`. This statement calls the function `matplotlib.pyplot.scatter` as a method of the object `ax`. Calling object `ax`'s `scatter` method with the parameters `x`, `y`, and `z` plots points on the figure. The resulting scatterplot is then assigned to the variable `plot`.

```
# import libraries
import matplotlib

# set coordinates
x = 0
y = 0
z = 0

# create figure
ax = matplotlib.pyplot.subplot(projection='3d')
```

```
# plot coordinates
plot = ax.scatter(x, y, z)
```

Printing Use the following statement at the end of the script to print the coordinates for this point:

```
print(x, y, z)
```

```
0 0 0
```

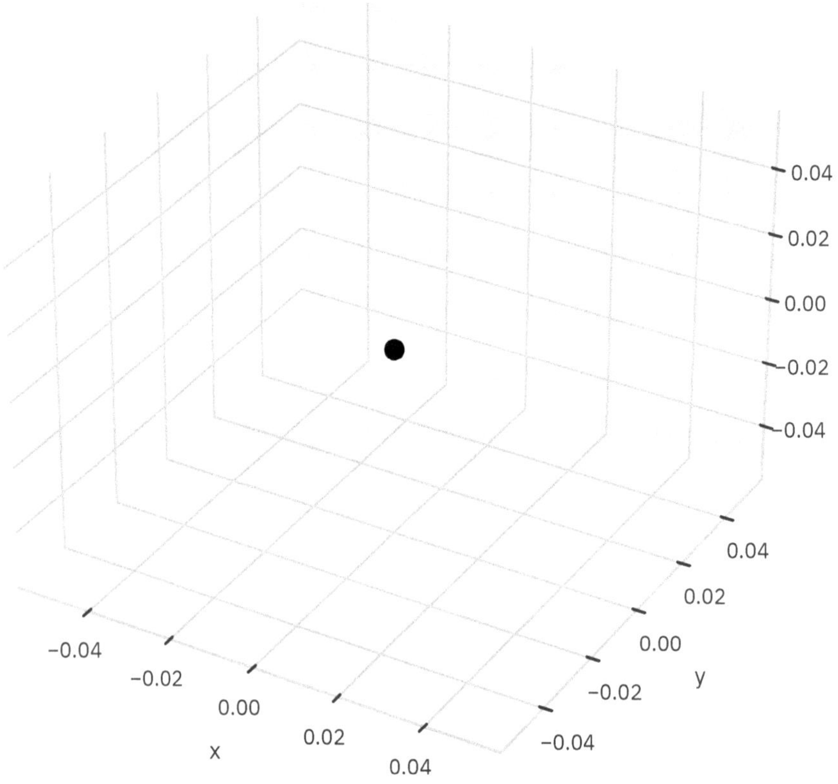

Figure 3.1: A single point at the origin of a Cartesian space

3.1.5 Hello Points

Use Python to generate a random population of points. This example uses two more lines of code than the previous example to generate a hundred random points instead of a single fixed point. NumPy – a library for performing mathematical operations on arrays [18] – will be used to generate random coordinates for each point.

Code: `hello-points.py`

Import Start by importing NumPy with the statement `import numpy as np` and Matplotlib with the statement `import matplotlib.pyplot as plt`. NumPy will be imported with the alias `np`. While in the previous example, the entire Matplotlib library was imported, this time only the `pyplot` module of the library will be imported with the alias `plt`. Aliases can make your code more concise and legible. For example the function `matplotlib.pyplot.subplot` can now be called with the statement `plt.subplot()` where `plt` is an alias for `matplotlib.pyplot`.

Figure Create an empty figure with `ax = matplotlib.pyplot.subplot(projection = '3d')`.

Randomness Create an instance of NumPy's pseudo-random number generator `numpy.random.default_rng()` and assign this object to the variable `rng`. Assign the number of points you want to the variable `n`. Then use the function `numpy.random.Generator.random` to generate an array of length `n` of random numbers between 0.0 and 1.0 for the x-coordinates of the points with the statement `x = rng.random(n)`. Generate arrays for the x- and y-coordinates the same way. Learn more about randomness in chapter 4.

Plot Plot the coordinates on the figure as a scatterplot with the statement `plot = ax.scatter(x, y, z, c='black', s=100)`. Set the color of the points with parameter `c` and the point size with parameter `s`. A named color such as `black` can be set as a string.

```
import numpy as np
import matplotlib.pyplot as plt

# create figure
ax = plt.subplot(projection='3d')

# instantiate random number generator
rng = np.random.default_rng()

# generate random x-, y-, and z-values
n = 100
x = rng.random(n)
y = rng.random(n)
z = rng.random(n)

# plot points
plot = ax.scatter(x, y, z, c='black', s=100)
```

Print Use a `for` loop to print the coordinates of each point. Define the header of the loop with the statement `for i in range(n):` where `n` is the number of points. In this statement, `range()` generates a sequence from 0 to `n`. The variable `i` will be the loop's position in that sequence and can be used as a counter. Define the body of the loop by indenting once and then typing `print(x[i], y[i], z[i])`. This will print the elements at position `i` in the arrays x, y, and z.

```
# print coordinates
for i in range(n):
    print(x[i], y[i], z[i])
```

```
0.8065465997163719 0.8834681099370751 0.2925904251013901
...
```

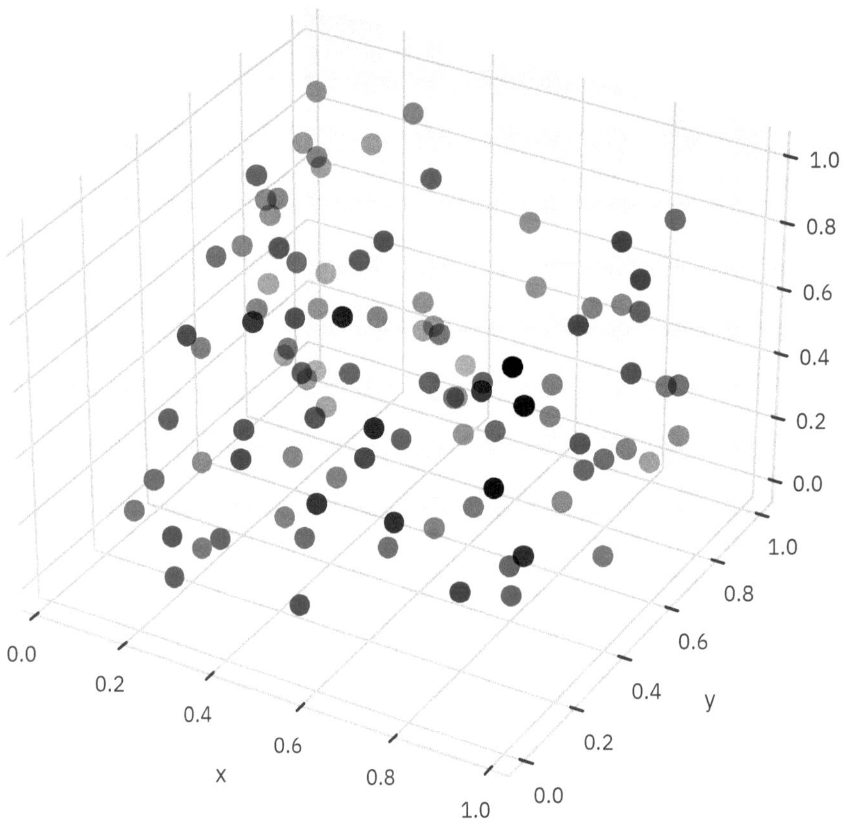

Figure 3.2: Random points

3.2 Rhino Scripting

With Python scripting in Rhino, designers can algorithmically generate geometry. While Rhino has a visual programming interface, scripting is useful for developing more complex algorithms. Rhino 8 supports scripting in CPython with the new ScriptEditor. The ScriptEditor can import modules from Python's standard library, community libraries like NumPy and SciPy, and the RhinoScriptSyntax package, which has functions for creating and transforming geometry in Rhino. To use the ScriptEditor, download and install Rhino 8.

Rhino uses the IronPython implementation of Python, which does not support libraries like NumPy and SciPy. This book uses Rhino's ScriptEditor instead for access to CPython's wider ecosystem of libraries and modules. The RhinoScriptSyntax package works with IronPython in Rhino and with CPython via the ScriptEditor. Guides to using Python in Rhino can be found at https://developer.rhino3d.com/guides/rhinopython including the Rhino Python Primer [24] and RhinoScriptSyntax in Python [22].

The ScriptEditor can also be used to create custom components in Grasshopper. Guides to using Python in Grasshopper include McNeel's "An Overview of the GhPython Component" [21] and Delft University of Technology's tutorial on "How to use GhPython" [16]. While these guides cover GhPython for IronPython scripting in Grasshopper, it is very similar to CPython scripting with the ScriptEditor, so their guidance will still be useful.

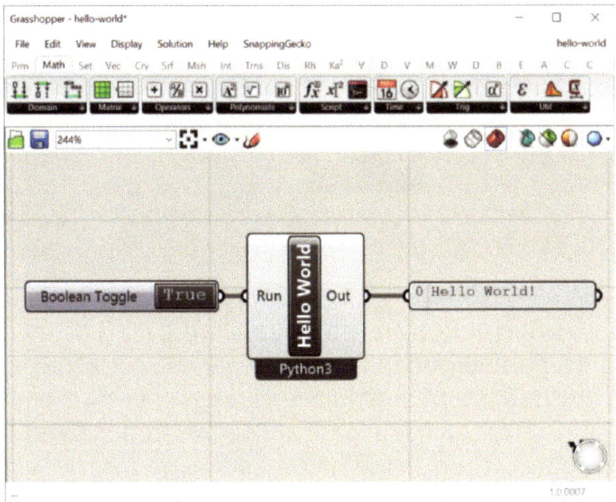

Figure 3.3: Grasshopper script component

Figure 3.4: ScriptEditor in Grasshopper

3.2.1 Hello World

Print a simple message using Python in the ScriptEditor. In Rhino's command line, run the command `ScriptEditor` to launch the editor. In the editor's file menu, create a new Python3 script. Then write a print statement and hit the `Run` button. The print statement will appear in the editor's terminal. To open the terminal, click on its tab at the bottom of the editor.

Code: `hello-world.gh`

```
ScriptEditor
```

```
print("Hello World")
```

Grasshopper Use the ScriptEditor to create a custom Grasshopper component that prints a simple message. The component will have a single input parameter and a single output parameter. The input will be a `Boolean Toggle` to run the script and the output will be a string, i.e. text.

Figure 3.5: Component name

Figure 3.6: Component icon

In Grasshopper place a `Script` component from the math tab on the canvas. Plug a `Boolean Toggle` into input `x`. Zoom in on the inputs, delete input `y`, and rename input `x` to `Run`. Double click on the component to open the ScriptEditor and write a new script. Write an `if` statement to test whether the variable `Run` is set to `True`. Indent once to start the body of the `if` statement and then write a `print()` statement. When the component runs, this statement will be printed in the ScriptEditor's terminal and in the component's standard output. Save and close the script. Plug the component's standard output parameter `out` into a `Panel` to display the print statement. Right click on the middle of the `Script` component to rename it `Hello World`. In Grasshopper double click on the `Boolean Toggle` to run the script and print your message on the `Panel`. The names of the Grasshopper component's parameters must match the names of the Python variables exactly.

```
if Run:
    print("Hello World!")
```

3.2.2 Hello Points

Generate a random population of points using Rhino's Script Editor. After generating a 3D array of points with NumPy, use RhinoScriptSyntax to create a point cloud in Rhino from these coordinates. The input for RhinoScriptSyntax's `AddPointCloud` method is a nested list – i.e. a list of points, each composed of a list of coordinates. If NumPy's `numpy.random.Generator.random` method is used to generate three separate lists of x-, y-, and z-coordinates, then these lists need to be interwoven into a single nested list. There are many ways to do this including appending in a `for` loop, parallel iteration, or stacking with NumPy.

Code: `random-points.py`

```
# requirements: numpy
import numpy as np
import rhinoscriptsyntax as rs

# instantiate random number generator
rng = np.random.default_rng()

# generate random x, y, and z values
n = 100
x = rng.random(n)
y = rng.random(n)
z = rng.random(n)

# set points from coordinates
points = np.column_stack((x, y, z))

# generate point cloud
rs.AddPointCloud(points)
```

Import Packages like NumPy need to be installed during the ScriptEditor's runtime. At the beginning of the script add a comment with the format `# requirements: numpy` to install a required package. The ScriptEditor will check these comments and install any required packages before running the script. Then specify the packages to import. For this example, import NumPy with `import numpy as np` and RhinoScriptSyntax with `import rhinoscriptsyntax as rs`.

Randomness Instantiate NumPy's random number generator with the statement `rng = np.random.default_rng()`, set the number of points with `n = 100`, generate random x-coordinates with `x = rng.random(n)`, and then do the same for the y- and z-coordinates.

Looping A `for` loop can be used to interweave the lists of coordinates into a nested list. First create an empty list named `points`. Then iterate through the coordinates with a `for` loop. During each iteration, append the x-, y-, and z-coordinates to the list `points` as a sublist.

```
points = []
for i in range(n):
    points.append((x[i], y[i], z[i]))
```

Parallel Iteration Python's built-in function `zip` can be used to interweave multiple lists by iterating through them in parallel.

```
points = list(zip(x, y, z))
```

Stacking The fastest, most computationally efficient way to interweave the arrays of coordinates is to stack them with NumPy's `numpy.column_stack` function.

```
points = np.column_stack((x ,y, z))
```

Points Generate points from the nested list of coordinates with the statement `rs.AddPoints(points)`. The input for RhinoScriptSyntax's `AddPoints` method is either a list of 3D point objects or a nested list of coordinates. This will generate points as geometry in Rhino that can be rendered in the viewports, transformed, exported, etc.

Point Cloud `Point clouds` are collections of points. Point clouds are a computationally efficient way to handle large populations of points. They can be generated from nested lists of coordinates with the RhinoScriptSyntax function `AddPointCloud`. Generate a point cloud from the nested list of coordinates with the statement `rs.AddPointCloud(points)`.

Grasshopper Use the ScriptEditor to develop a custom Grasshopper definition for generating random points. In Grasshopper place a `Script` component on the canvas. Set the parameters for the component by renaming input `x` to `Run`, input `y` to `Count`, and output `a` to `Cloud`. Plug a `Boolean Toggle` into the `Run` parameter and plug a `Number Slider` into the `Count` parameter. Right click on the middle of the component to rename it to `Random Points`.

Double click on the `Script` component to start writing the script. Adapt the script `random-points.py` from section 3.2.2 for Grasshopper. To create geometry using Python in Grasshopper, set the script context to Rhino. To do this, import Rhino and then ScriptContext with **`import scriptcontext as sc`** and then set the context to Rhino with `sc.doc = Rhino.RhinoDoc.ActiveDoc`. After importing the required modules, nest the rest of the script inside an `if` statement that tests whether to run the script. Set the variable `n` to the input parameter `Count` and cast this as an integer with the statement `n = int(Count)`. Finally assign the point cloud generated by RhinoScriptSyntax's `AddPointCloud` method to the variable `Cloud` with the statement `Cloud = rs.AddPointCloud(points)`.

Code: `random-points.gh`

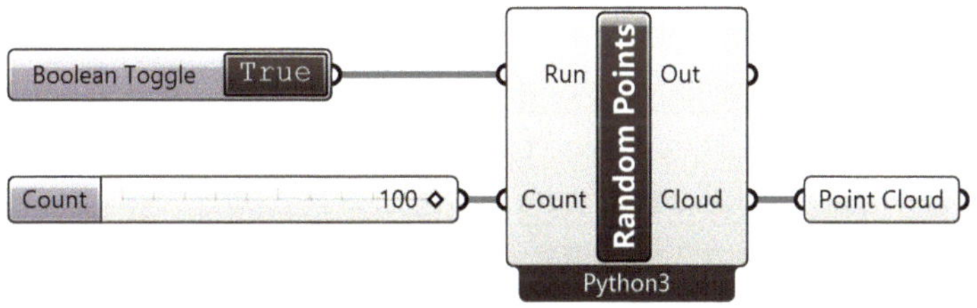

```python
# requirements: numpy
import numpy as np
import Rhino
import rhinoscriptsyntax as rs
import scriptcontext as sc
sc.doc = Rhino.RhinoDoc.ActiveDoc

if Run:

    # instantiate random number generator
    rng = np.random.default_rng()

    # generate random x-, y-, and z-values
    n = int(Count)
    x = rng.random(n)
    y = rng.random(n)
    z = rng.random(n)

    # set points from coordinates
    points = np.column_stack((x, y, z))

    # generate point cloud
    Cloud = rs.AddPointCloud(points)
```

References

[1] [Software] Anaconda, *Miniconda* version 23.5.0, 2023. lic: BSD 3-Clause License. url: https://docs.conda.io/en/latest/miniconda.html, vcs: https://repo.anaconda.com/miniconda.

[2] [Software] compas-dev, *COMPAS: A computational framework for collaboration and research in Architecture, Engineering, Fabrication, and Construction* version 1.17.5, 2023. lic: MIT License. doi: 10.5281/zenodo.7646467, url: https://compas.dev, vcs: https://github.com/compas-dev/compas.

[3] [Software] Conda Contributors, *Conda: A system-level, binary package and environment manager running on all major operating systems and platforms.* Version 23.5.2, 2023. lic: BSD 3-Clause License. url: https://conda.org/, vcs: https://github.com/conda/conda.

[4] [Software] Spyder Project Contributors, *Spyder: the Scientific Python Development Environment* version 5.4.3, 2023. lic: MIT License. url: https://www.spyder-ide.org, vcs: https://github.com/spyder-ide/spyder.

[5] [Software] NumPy Developers, *NumPy* version 1.25.0, 2023. lic: BSD 3-Clause License. url: https://numpy.org, vcs: https://github.com/numpy/numpy.

[6] [Software] Python Software Foundation, *Python* version 3.11.4, 2023. lic: Python Software Foundation License. url: https://www.python.org/, vcs: https://github.com/python.

[7] [Software] GDAL/OGR Contributors, *GDAL/OGR Geospatial Data Abstraction software Library* version 3.7.1, 2023. Open Source Geospatial Foundation. lic: MIT License. doi: 10.5281/zenodo.8144520, url: https://gdal.org, vcs: https://github.com/OSGeo/gdal.

[8] [Software] Sean Gillies, Casper van der Wel, Joris Van den Bossche, Mike W. Taves, Joshua Arnott, Brendan C. Ward, et al., *Shapely* version 2.0.1, 2023. lic: BSD 3-Clause License. doi: 10.5281/zenodo.7428463, url: https://shapely.readthedocs.io, vcs: https://github.com/shapely/shapely.

[9] [Software] Kelsey Jordahl, Joris Van den Bossche, Martin Fleischmann, Jacob Wasserman, James McBride, Jeffrey Gerard, Jeff Tratner, Matthew Perry, Adrian Garcia Badaracco, Carson Farmer, Geir Arne Hjelle, Alan D. Snow, Micah Cochran, Sean Gillies, Lucas Culbertson, Matt Bartos, Nick Eubank, maxalbert, Aleksey Bilogur, Sergio Rey, Christopher Ren, Dani Arribas-Bel, Leah Wasser, Levi John Wolf, Martin Journois, Joshua Wilson, Adam Greenhall, Chris Holdgraf, Filipe, and François Leblanc, *GeoPandas* version 0.13.2, 2023. lic: BSD 3-Clause License. doi: 10.5281/zenodo.8009629, url: https://geopandas.org, vcs: https://github.com/geopandas/geopandas.

[10] [Software] Martín Abadi et al., *TensorFlow: Large-Scale Machine Learning on Heterogeneous Systems* version 2.13.0, 2023. lic: Apache License 2.0. doi: 10.5281/zenodo.8118033, url: https://www.tensorflow.org, vcs: https://github.com/tensorflow/tensorflow.

[11] [Software] scikit-learn Developers, *scikit-learn* version 1.3.0, 2023. lic: BSD 3-Clause License. doi: 10.5281/zenodo.8098905, url: https://scikit-learn.org, vcs: https://github.com/scikit-learn/scikit-learn.

[12] [Software] SciPy Developers, *SciPy* version 1.11.1, 2023. lic: BSD 3-Clause License. url: https://scipy.org, vcs: https://github.com/scipy/scipy.

[13] [Software] Matplotlib Development Team, *Matplotlib* version 3.7.2, 2023. lic: Matplotlib License. url: https://matplotlib.org, vcs: https://github.com/matplotlib/matplotlib.

[14] Python Software Foundation. *What is Python? Executive Summary*. 2022. url: https://www.python.org/doc/essays/blurb.

[15] [Software] Michael Waskom, *Seaborn: statistical data visualization* version 0.12.2, 2022. lic: BSD 3-Clause License. url: https://seaborn.pydata.org, vcs: https://github.com/mwaskom/seaborn.

[16] Delft University of Technology. *Grasshopper "Good Practice"*. TOI-Pedia. 2021. url: http://wiki.bk.tudelft.nl/toi-pedia/Grasshopper (visited on 09/04/2022).

[17] Michael L. Waskom. "seaborn: statistical data visualization". In: *Journal of Open Source Software* 6.60 (2021), page 3021. doi: 10.21105/joss.03021.

[18] Charles R. Harris, K. Jarrod Millman, Stéfan J. van der Walt, Ralf Gommers, Pauli Virtanen, David Cournapeau, Eric Wieser, Julian Taylor, Sebastian Berg, Nathaniel J. Smith, Robert Kern, Matti Picus, Stephan Hoyer, Marten H. van Kerkwijk, Matthew Brett, Allan Haldane, Jaime Fernández del Río, Mark Wiebe, Pearu Peterson, Pierre Gérard-Marchant, Kevin Sheppard, Tyler Reddy, Warren Weckesser, Hameer Abbasi, Christoph Gohlke, and Travis E. Oliphant. "Array programming with NumPy". In: *Nature* 585.7825 (Sept. 2020), pages 357–362. doi: 10.1038/s41586-020-2649-2.

[19] [Software] The Pandas Development Team, *Pandas* version 2.0.3, 2020. lic: BSD 3-Clause License. doi: 10.5281/zenodo.3509134, url: https://pandas.pydata.org, vcs: https://github.com/pandas-dev/pandas/tree/v2.0.3.

[20] Pauli Virtanen, Ralf Gommers, Travis E. Oliphant, Matt Haberland, Tyler Reddy, David Cournapeau, Evgeni Burovski, Pearu Peterson, Warren Weckesser, Jonathan Bright, Stéfan J. van der Walt, Matthew Brett, Joshua Wilson, K. Jarrod Millman, Nikolay Mayorov, Andrew R. J. Nelson, Eric Jones, Robert Kern, Eric Larson, C. J. Carey, İlhan Polat, Yu Feng, Eric W. Moore, Jake VanderPlas, Denis Laxalde, Josef Perktold, Robert Cimrman, Ian Henriksen, E. A. Quintero, Charles R. Harris, Anne M. Archibald, Antônio H. Ribeiro, Fabian Pedregosa, Paul van Mulbregt, and SciPy 1.0 Contributors. "SciPy 1.0: Fundamental Algorithms for Scientific Computing in Python". In: *Nature Methods* 17 (2020), pages 261–272. doi: 10.1038/s41592-019-0686-2.

[21] Scott Davidson. *An Overview of the GhPython Component*. Robert McNeel & Associates, 2019. url: https://developer.rhino3d.com/guides/rhinopython/ghpython-component.

[22] Dale Fugier. *RhinoScriptSyntax in Python*. Robert McNeel & Associates, 2019. url: https://developer.rhino3d.com/guides/rhinopython/python-rhinoscriptsyntax-introduction.

[23] F. Pedregosa, G. Varoquaux, A. Gramfort, V. Michel, B. Thirion, O. Grisel, M. Blondel, P. Prettenhofer, R. Weiss, V. Dubourg, J. Vanderplas, A. Passos, D. Cournapeau, M. Brucher, M. Perrot, and E. Duchesnay. "Scikit-learn: Machine Learning in Python". In: *Journal of Machine Learning Research* 12 (2011), pages 2825–2830.

[24] Skylar Tibbits, Arthur van der Harten, and Steve Baer. *Rhino Python Primer*. Robert McNeel & Associates, 2011. url: https://developer.rhino3d.com/guides/rhinopython/primer-101.

[25] Wes McKinney. "Data Structures for Statistical Computing in Python". In: *Proceedings of the 9th Python in Science Conference*. Edited by Stéfan van der Walt and Jarrod Millman. 2010, pages 56–61. doi: 10.25080/Majora-92bf1922-00a.

[26] John D. Hunter. "Matplotlib: A 2D Graphics Environment". In: *Computing in Science and Engineering*. 9.3 (June 2007), pages 90–95. issn: 1521-9615. doi: 10.1109/MCSE.2007.55.

4 Randomness

4.1 Stochastic Processes

Natural landscapes are irregular as they are shaped in part by stochastic processes. There is an element of chance in any landscape; seeds carried by birds or the wind, the changing structure of plants as they grow, and the vagaries of weather all introduce uncertainty. While trees can be planted in regular rows or grids for an aesthetic of repetition and order, a degree of irregularity, of unpredictability is needed when designing naturalistic landscapes. In computational design, randomness is an easy way to introduce a degree of unpredictability into a design. Used creatively a carefully measured dose of randomness can disrupt the predictability of deterministic algorithms and create aesthetic tension between order and disorder.

4.2 Randomness in Grasshopper

Create 1- and 2-dimensional arrays of random numbers with Grasshopper using the definition random.gh as a guide.

Code: `random.gh`

4.2.1 1D Random Array

Use the `Random` component – with a Schrödinger's cat icon – to generate a list of random numbers. Use number sliders to set the top of the domain for the numeric range, the number of random values to generate, and the seed of the random number generator. Use a panel to quickly visualize the resulting list of random numbers. This example returns 10 random floating point numbers between 0 and 1. Because this pseudo-random number generator requires a set seed, the result will be the same every time for a given seed allowing for reproducible computations.

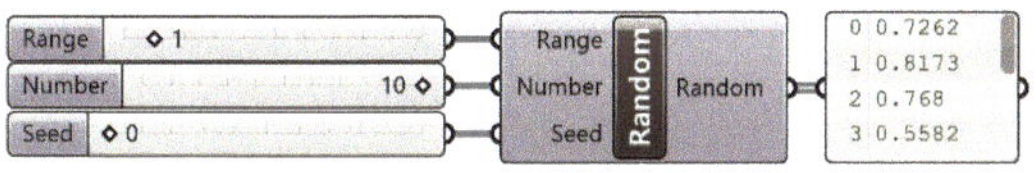

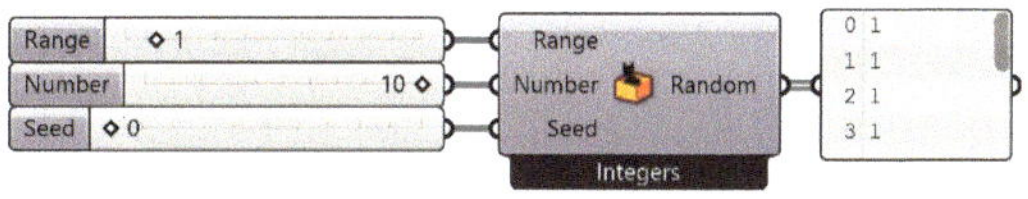


```
0.7262 0.8173 0.768 0.5582 0.206 0.5589 0.906 0.4422 0.9775 0.2737
```

Right click on the `Random` component to set integer mode.

```
1 1 1 1 0 1 1 0 1 0
```

4.2.2 2D Random Array

To generate and visualize a 2-dimensional array of random numbers in Grasshopper, construct a grid, generate a random value for each grid cell, and then color each cell relative to its random value. Start by constructing a grid with the `Square` component. Flatten the output data tree of cells into a list. Convert each grid cell into a surface with the `Boundary Surfaces` component. Then find how many cells are in the grid with the `List Length` component. Generate that many random numbers with the `Random` component. Plug the list of random numbers into a `Gradient` component and right click to set a default color map or double click to design custom color map. Place a `Custom Preview` component, plug the surfaces in as the input geometry, and plug the color gradient in as the input material. The range of the `Random` component and the lower and upper limits of the `Gradient` should match. For simplicity's sake in this example both use the default domain of 0 to 1. This array can also be visualized as a matrix with the `Construct Matrix` and `Display Matrix` components.

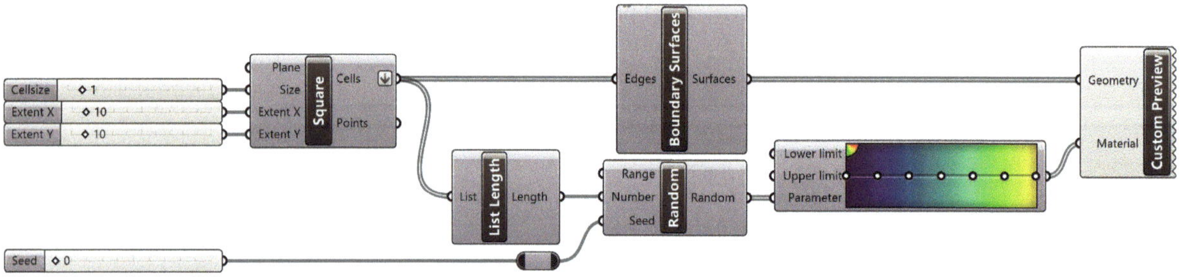

4.3 Randomness in Python

NumPy's `numpy.random.Generator.random` method generates random floating point numbers that are drawn from a continuous uniform distribution. In a continuous uniform distribution each value within the interval – here, from 0 and 1 (including 0, but excluding 1) – is equally likely to occur. NumPy's default algorithmic pseudo-random number generator – PCG64 – can draw random samples from a wide range of other probability distributions including uniform, standard normal, exponential, binomial, logistic, Gaussian, Poisson, Laplace, and Pareto distributions [1].

4.3.1 1D Random Array

Floating Point Numbers Create a 1-dimensional array of random floating point numbers drawn from a continuous uniform distribution with NumPy using the script `random-floats.py` as a guide. First import Numpy and then initialize its default random number generator with `numpy.random.default_rng` and assign it to the variable `rng`. For reproducible results, set a seed such as 12345 for the generator with: `rng = np.random.default_rng(12345)`. Then generate random floating point numbers with the method `numpy.random.Generator.random`. Set the number of random samples to return.

Code: `random-floats.py`

```
# import libraries
import numpy as np

# generate random array
rng = np.random.default_rng()
z = rng.random(1000)
print(z)
```

```
[0.22263676 0.65712042 0.6670191 0.8521939 0.06159513 0.27130182 0.68210952 ...]
```

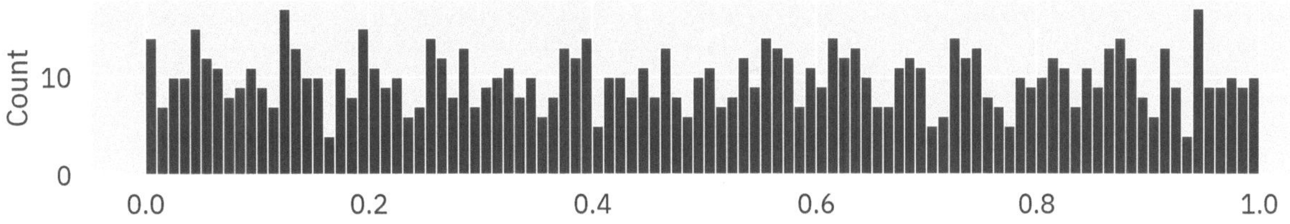

Integers To return an array of integers, use the `numpy.random.Generator.integers` method. Set the bottom of the range, the top of the range, and the number of samples to return.

Code: `random-integers.py`

```
z = rng.integers(1, 100, 1000)
```

```
[97 97 98 38 17 92 11 26 14 32 54 12 79 24 78 32 79 96 95 27 56 44 26 61 89...]
```

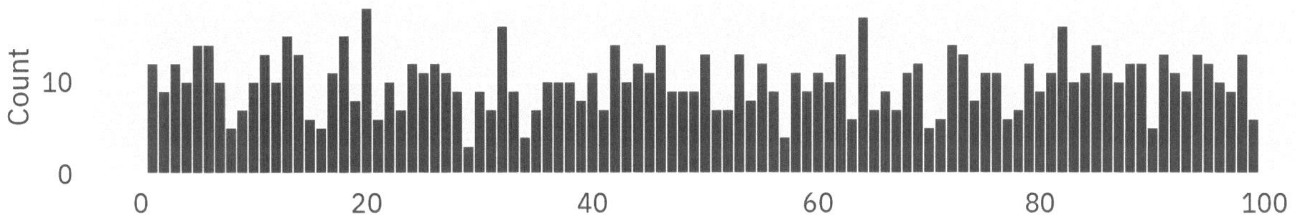

Normal Distribution To draw random samples from the Gaussian, i.e. normal, distribution, use the method `numpy.random.Generator.normal`. Set the mean of the distribution, the standard deviation of the distribution, and the number of samples to return.

Code: `random-normal.py`

```
z = rng.normal(0, 10, 1000)
```

```
[1.58869994e+01  7.33091465e+00 -6.57176781e+00  1.66879605e+01...]
```

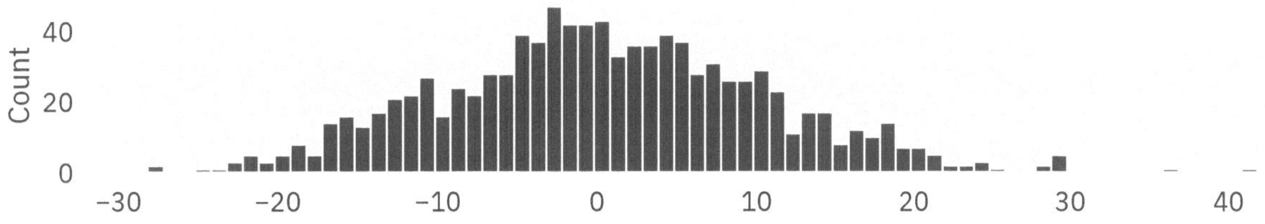

Laplace Distribution To draw random samples from the Laplace, i.e. double exponential, distribution, use the method `numpy.random.Generator.laplace`. Set the peak of the distribution, the exponential decay, and the number of samples to return.

```
z = rng.laplace(0, 10, 1000)
```

```
[3.06614472e+01 -2.73921886e+00  1.87400564e+01 -6.47438239e+00...]
```

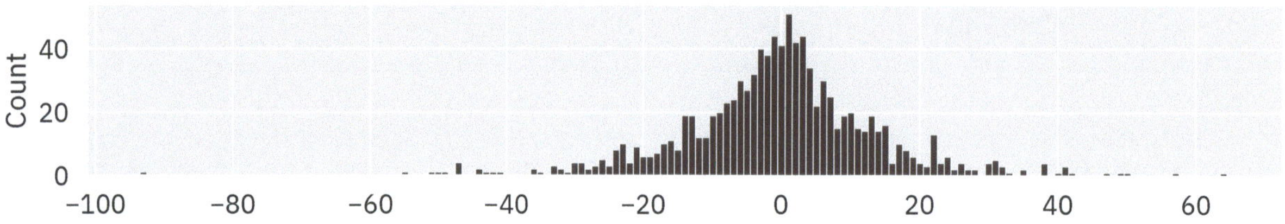

4.3.2 2D Random Array

Create a 2-dimensional array of integers with the method `numpy.random.Generator.integers`. Set the bottom of the range, the top of the range, and the size of the array to return. To return a 10 by 10 array, set the size to `(10, 10)`. Plot the array as a heatmap with `seaborn.heatmap`. Set the color map to a perceptually uniform color gradient such viridis, inferno, or magma (appendix B).

Code: `random-2d.py`

```
# import libraries
import seaborn as sns
from numpy.random import default_rng

# set theme
sns.set_theme(style="white")

# generate random matrix
rng = default_rng()
z = rng.integers(0, 10, (10, 10))

# generate heatmap
sns.heatmap(z, cmap='viridis')
```

References

[1] Melissa E. O'Neill. *PCG: A Family of Simple Fast Space-Efficient Statistically Good Algorithms for Random Number Generation.* Technical report HMC-CS-2014-0905. Claremont, CA: Harvey Mudd College, Sept. 2014. url: https://www.cs.hmc.edu/tr/hmc-cs-2014-0905.pdf.

5 Random Paving

5.1 Paving Patterns

Generate random paving patterns using the Grasshopper definition `random-paving.gh` as a guide. First create a rectangular grid of pavers in a stacked bond pattern. Then shift alternating rows of pavers to form a running bond pattern. Apply a random color map to the pavers. Finally use a curve attractor to create a more structured, intentional color scheme for the paving. Useful plugins for paving include Parakeet [2] and Land Kit [1].

Code: `random-paving.gh`

5.2 Stacked Bond

First layout a grid of pavers in a stacked bond pattern. Start by generating a grid with the `Rectangular` component. Then generate a box at each point in the grid with the `Domain Box` component. Set the base of the boxes to the points in the grid and set the size of the boxes in the x-, y-, and z-directions. Plug the x- and y-size of the grid into `Domain Box`'s x and y inputs. Optionally, subtract an offset factor to create space for mortar around the pavers.

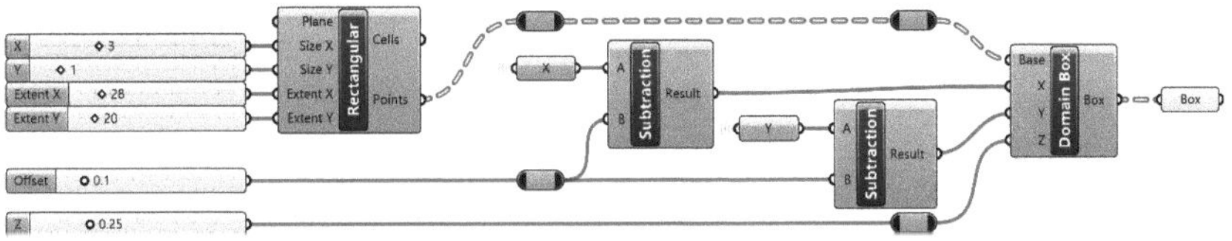

5.3 Running Bond

Shift alternating rows of pavers to create a running bond pattern. Split the data tree of pavers into alternating rows with the `Dispatch` component. Then move the pavers in one of the resulting data trees along the x-axis by half the length of the paver with the `Move` component. Recombine the data trees of pavers with the `Merge` component and flatten the result.

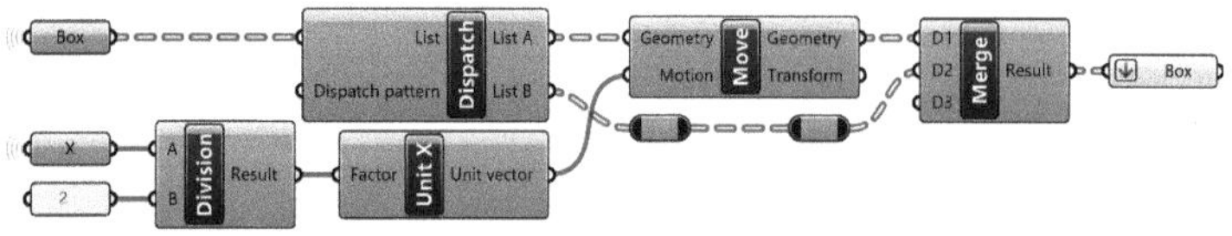

5.4 Random Colors

Assign random colors to each paver. First find the number of pavers with the `List Length` component. Generate a sequence of random numbers of that length with the `Random` component. Plug the list of random numbers into a `Gradient` as the parameter input and right click to set the default grayscale color map. Add a `Custom Preview` and set the input geometry to the list of pavers and the material to the color gradient.

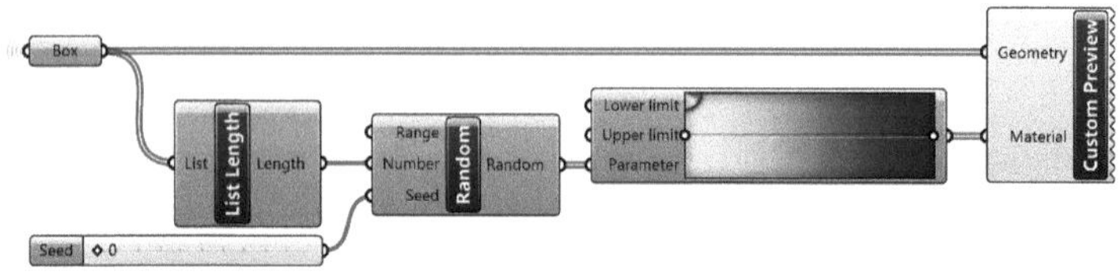

5.5 Paving Attractor

Draw a curve attractor to give some structure to the random field of color. The lighter paving will mark a route across the plaza. First find the centroid of each paver with the `Area` component. Then draw a curve in Rhino, set it as a curve parameter, and divide it into a sequence of points using the `Divide Distance` component. Find the distance between the centroids and the points along the curve with the `Closest Points` component. Remap the distance between the centroids and the attractor points to a domain from 0 to 1 with the `Remap Numbers` component. Generate a list of the same length of random numbers with the `Random` component. Multiply the remapped list of distances by the list of random numbers and by a scale factor with the `Multiplication` component. Use the resulting list as the parameter input for a `Gradient`. Place a `Custom Preview` and plug the pavers in as the input geometry and the gradient in as the input material.

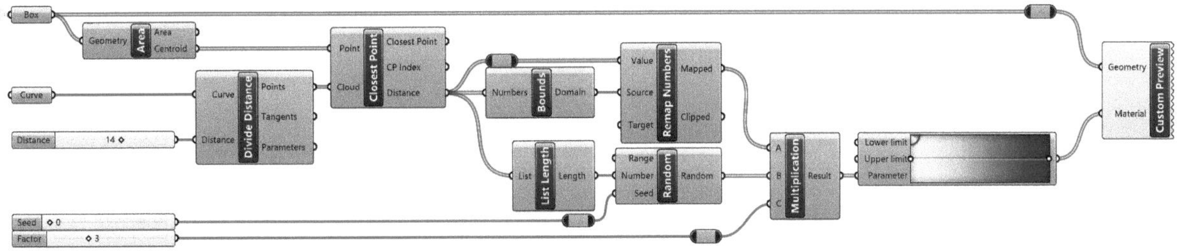

References

[1] [Software] Landau Design+Technology, *Land Kit* version 1.2, 2022. url: https://www.landkit.design/, vcs: https://www.food4rhino.com/en/app/land-kit.

[2] [Software] Esmaeil Mottaghi and Arman KhalilBeigi, *Parakeet* version 0.9.3, 2022. vcs: https://www.food4rhino.com/en/app/parakeet.

6 Tessellations

6.1 Cellular Patterns

In a Voronoi diagram – also known as a Dirichlet tessellation – a plane is divided into cells based on the location of a set of points [4, 3]. Each point is assigned the region of space that is closer to it than to any other point. These points may be data or random seeds. Voronoi diagrams have been used to map phenomena in the spatial sciences and to generate cellular, space-filling geometry in design [2]. Voronoi diagrams with random seeds can be used by landscape architects to divide space into cellular patterns for plant communities, paving, and so forth. A Voronoi diagram could be used, for example, to model a rockery like Terrain Work's Pop Rocks art installation [1].

Use a Voronoi diagram to generate a random cluster of rocks in a cellular pattern. First generate a Voronoi diagram from a random set of points in a planar region. Then extrude the cells of the Voronoi diagram by a factor proportional to their area. Color the cells so that their brightness is proportional to their area.

Code: `cellular-rockery.gh`

6.2 Voronoi Diagram

Generate a Voronoi diagram for a random set of points. First model a `Rectangle` and set its x- and y-dimensions with number sliders. Then generate a random set of points with the `Populate 2D` component. Set the input region to the rectangle and use number sliders to set the input count and seed. Generate a cellular network with a `Voronoi` component. Set the input points to the random population and the input boundary to the rectangle.

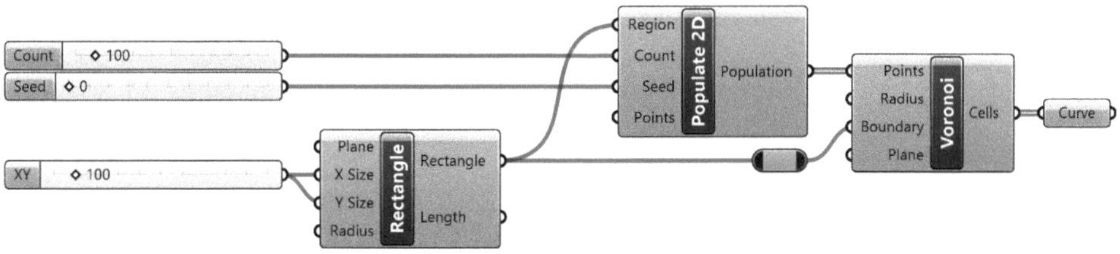

6.3 Extrusion

Extrude the cells relative to their area. Model the cells as planar surfaces with the `Boundary Surfaces` component. Then `Extrude` the planar surfaces into solid volumes. To set the direction for and magnitude of extrusion, first calculate the area of each cell with the `Area` component. Then calculate the domain of the list of areas with the `Bounds` component. Remap the list of areas with the Remap Numbers component. Set the input value to the list of areas, the source to the domain of the list of areas, and the target to a new domain such as 2 To 10. Set the remapped list as the input factor for the Unit Z vector. Set the resulting vector as the direction for the Extrude component.

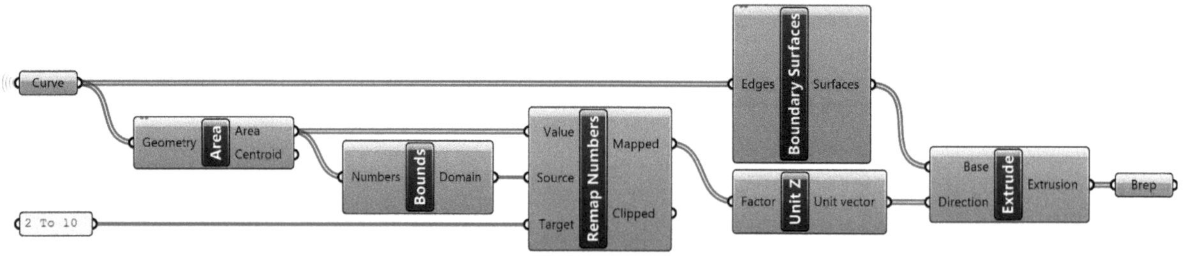

6.4 Color Gradient

Apply a color gradient to the rockery with brightness relative to the area of the stones. First calculate the domain of the list of areas with the Bounds component. Then remap the list of areas to a domain from 0 to 1 with the Remap Numbers component. Set the input value to the list of areas, the source to domain of the list of areas, and leave the input target as the default, i.e. 0 To 1. Use the remapped list as the input parameter for a Gradient and set a color map. Add a Custom Preview component and set the input geometry to the list of extruded cells and set the input material to the color gradient.

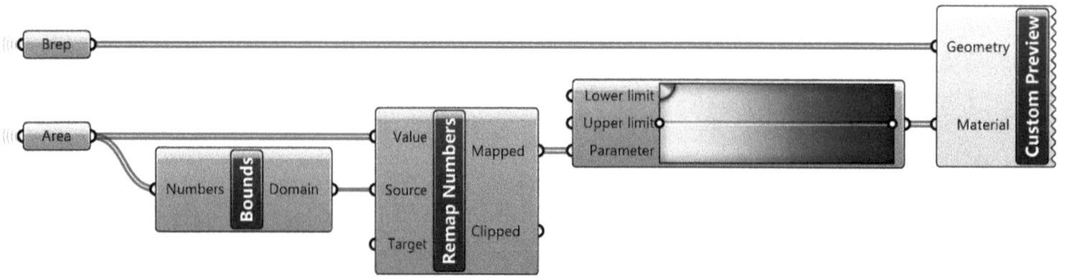

References

[1] Terrain Work. *Pop Rocks*. 2019. url: https://www.terrainwork.com/work#/pop-rocks/ (visited on 08/04/2022).

[2] Mitchell Whitelaw. "Space filling And Self-Constraint". In: *Architectural Theory Review* 15.2 (2010), pages 157–165. doi: 10.1080/13264826.2010.495451.

[3] Adrian Bowyer. "Computing Dirichlet tessellations". In: *The Computer Journal* 24.2 (1981), pages 162–166. doi: 10.1093/comjnl/24.2.162.

[4] Georges Voronoï. "Nouvelles applications des paramètres continus à la théorie des formes quadratiques. Premier mémoire. Sur quelques propriétés des formes quadratiques positives parfaites". In: *Journal für die Reine und Angewandte Mathematik* 133 (1908), pages 97–178.

7 Point Clouds

7.1 Point Cloud Data

A point cloud is a collection of points. Each point is defined by spatial data – by x-, y-, and z-coordinates – and may have other attributes such as color. Points are a fundamental type of geometry and can be created and used in many different ways. They can be generated algorithmically or collected with global navigation satellite systems, photogrammetry, laser scanning, or other depth sensors. Since they are the most basic type of geometry, points are often used to model other types of geometry such as rasters or meshes. While long used as records, point clouds are beginning to emerge as a design medium [25] with a unique aesthetic. In this book point clouds will be used to model and visualize landscapes.

7.1.1 Creation

As demonstrated in previous chapters, point clouds can be generated algorithmically, but they can also be collected from the physical environment. There are many different technologies for sensing the physical environment. Given the complexity of our environments, multiple technologies are often used in tandem to capture different aspects of a space. While landscapes can be captured from above as point clouds with airborne laser scanning or structure from motion photogrammetry, their forest understory can be captured in finer detail with terrestrial laser scanning [39, 30, 19]. While a city can be captured from above with airborne laser scanning, ground-based sensors such as handheld, tripod-mounted, or vehicle-mounted laser scanners are needed to capture the details of interiors and streetscapes.

Algorithmic Computational design processes for algorithmically generating point clouds include randomness, noise, and physics simulations. While point clouds collected using lidar or photogrammetry represent the existing physical environment, algorithmically generated point clouds can be a creative medium – a way to model new ideas for a site.

Lidar Laser scanning is also known as lidar – light detection and ranging. Some lidar systems calculate distance based on the time-of-flight of a pulse of light, while others calculate distance based on the phase shift of a continuous beam of light. Time-of-flight lidar systems emit laser pulses that reflect off of surfaces in the environment and then return to the scanner. Distance (d) is calculated as function of time (t) times the speed of light (c) divided by 2. Position is then calculated from the location and orientation of the scanner and the distance to the target.

$$d = \frac{t \cdot c}{2}$$ **Equation** 7.1

Phase-shift lidar systems calculate distance from the waveform of a continuous beam of light. Since phase-shift systems are relatively fast, but short range, they are often used to scan buildings, interiors, and objects, while time-of-flight lidar systems with longer range and slower acquisition rates are used for topographic surveying and airborne laser scanning. Contemporary phase-shift systems like the Faro Focus Premium can capture up to 2 million points per second at a max range of 150 meters [8], while pulsed time-of-flight systems like the RIEGL VZ-2000i can capture between 500,000 points per second at a max range of 600 meters and 21,000 points per second at a max range of 2,500 meters [14]. The resulting point clouds can have millions or even billions of points.

In airborne lidar, differential global navigation satellite systems and inertial navigation systems are used in conjunction with laser scanners to calculate the position of scanned points in geographic space. While lidar systems mounted on manned aircraft can capture entire regions at sub-meter resolution and have been used to map much of the contiguous United States, drones with lidar can capture landscapes at even higher, sub-centimeter resolution (Figure 7.1).

Terrestrial Laser Scanner
Faro Focus

Drone Lidar
DJI Matrice 300 RTK + L1 Lidar

Figure 7.1: Lidar platforms

Structure from motion In structure from motion photogrammetry, 3-dimensional scenes are reconstructed from overlapping photographs taken from different views [54]. A photograph is flat; it is a 2-dimensional projection of a 3-dimensional scene. By looking at a photograph, however, we can estimate the depth of the scene because each of our eyes sees corresponding features from a slightly different view. Structure from motion does this computationally by triangulating between corresponding points on photographs. Feature detection algorithms such as the scale-invariant feature transform [50] are used to match corresponding features in different photographs. These corresponding points are then used to estimate the calibration and pose of the cameras and the position of 3-dimensional points in the scene. Then the position of the points and poses of the cameras are iteratively refined through bundle adjustment [44].

Typically a structured series of photographs from a camera in motion – such as a camera mounted on a drone – is used. With structured imagery, camera calibration, global navigation satellite system, and inertial measurement unit data can be used to simplify the calculations. However, unstructured imagery can also be used, albeit with messier, incomplete results. For example, crowd-sourced photographs from massive online repositories have been used to reconstruct cities [42, 38].

A systematic flight plan and ground control points are needed to accurately map a landscape using structure from motion photogrammetry with drone imagery. The flight plan should be systematic to ensure complete coverage with high overlap between photographs. Ground control points should be established before the flight. These are surveyed markers – often spray painted – that are visible from the air. The coordinates of ground control points can be collected quickly using real-time kinematic positioning with a base station and rover. They are used to optimize the alignment of cameras in structure from motion processing. Using ground control points can correct geometric distortions and locate the scene precisely in geographic space. Software for structure from motion photogrammetry include Metashape [1], Pix4D [11], and OpenDroneMap [5].

While it has limitations, structure from motion photogrammetry can generate dense, highly accurate point clouds at much less cost than lidar. While lidar systems are expensive, structure from motion photogrammetry can be done with relatively inexpensive cameras. Lidar processing and structure from motion are both computationally intensive. While lidar can partially penetrate vegetation – potentially recording the structure of trees and the ground beneath – structure from motion cannot. When done well structure from motion can achieve accuracy that is almost comparable to terrestrial laser scanning, but there are many possible sources of error including wind, changing light conditions, lens distortion, blurry or occluded photographs, and poorly marked ground control points [19].

7.1.2 Data Structure

Point clouds can be saved in a wide variety of formats including plain text files and binary point cloud files. They can also be saved as binary 3D geometry formats such as `.3dm` Rhino models. The data for this book are available in `.xyz`, `.e57`, `.laz`, `.pcd`, and `.3dm` formats.

Plain Text The simplest format for a point cloud is a plain text file. Point clouds saved as plain text files may have file extensions such as `.txt`, `.csv`, `.xyz`, or `.pts`. In a text file encoding a point cloud, each point is recorded on a new line as a list of coordinates separated by a delimiter such as a comma: `x,y,z` . These examples are random populations of points generated algorithmically with the scripts `random-point-cloud.py` and `color-point-cloud.py` (Figure 7.2–7.3).

Code: `random-point-cloud.py` & `color-point-cloud.py`

```
9,5,5
9,1,4
9,7,2
3,2,6
1,7,8
9,3,8
1,7,6
2,9,8
1,9,5
3,2,6
```

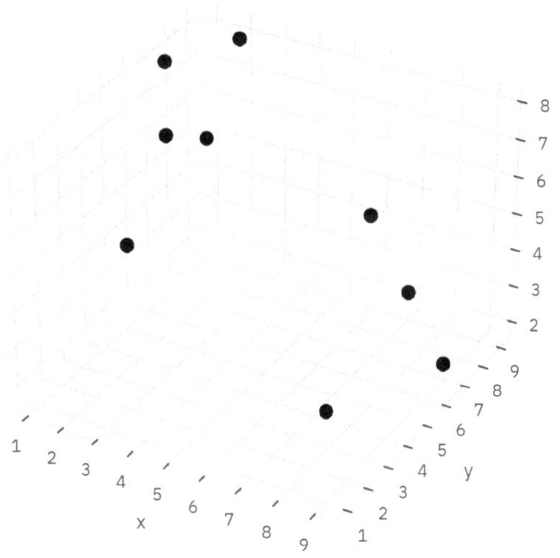

Figure 7.2: A randomly generated point cloud

If the point cloud also has color attributes, then these would be listed after the coordinates: `x,y,z,r,g,b`.

```
9,9,6,0.20803,0.718701,0.472873
9,5,2,0.244972,0.287675,0.53726
9,1,2,0.244972,0.287675,0.53726
3,7,6,0.20803,0.718701,0.472873
1,1,5,0.119699,0.61849,0.536347
9,4,3,0.190631,0.407061,0.556089
1,5,4,0.147607,0.511733,0.557049
2,3,1,0.281412,0.155834,0.469201
1,6,4,0.147607,0.511733,0.557049
3,9,4,0.147607,0.511733,0.557049
...
```

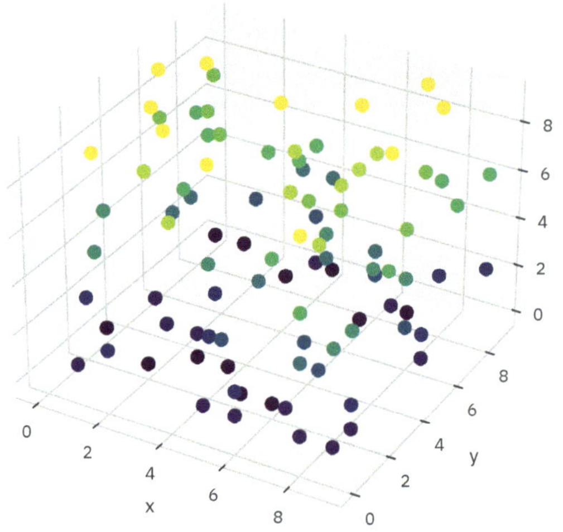

Figure 7.3: A randomly generated point cloud colored by elevation

Binary There are two widely used open formats for point cloud data. The `.las` specification is used extensively by the aerial sensing community, while the more general `.e57` standard is often used by the terrestrial sensing community. The older `.las` specification was developed by the American Society for Photogrammetry and Remote Sensing (ASPRS) to store and exchange airborne lidar data [37, 32, 24]. This binary format has a header, variable and extended variable length records, and point data records. The variable length records store information about the point cloud such as projection information, statistics, and metadata. The point data records include x-, y-, and z-coordinates, intensity, returns, classifications, and color. Points can be classified – either manually or algorithmically – as ground, low, medium, and high vegetation, buildings, water, and so forth. The open source `.laz` format is a compressed version of this specification that is 80–93 percent smaller [26, 13] and is used by agencies such as the United States Geological Survey to distribute airborne lidar data. The `.e57` format was developed by ASTM as an open standard for 3D imaging data from any depth sensor [40]. This format is composed of a binary header that describes the dataset, a hierarchical tree structure, and compressed binary data. It can store multiple point clouds and images in different coordinate systems as a single dataset.

7.2 Design

Point clouds are emerging as a creative medium in landscape architecture [25]. Designers use point clouds not only as records of existing physical spaces (Figure 7.4), but also as a way to model new spaces. Landscape architects at the Swiss Federal Institute of Technology in Zürich are exploring how georeferenced point clouds can be acquired, prepared, disassembled, transformed, and reassembled to model new designs for landscapes [29, 22, 17]. To regrade a landscape, bare earth points can be segmented from a georeferenced point cloud, transformed, and then merged back into the point cloud. Trees and other elements can be segmented out of the point cloud and compiled into a design library. When more detail is needed, terrestrial laser scanning can be used to compile a high resolution library of design elements. These elements can then be distributed across the georeferenced point cloud of the landscape. This book uses point clouds to algorithmically design landscapes. It explores algorithmic approaches to generating terrain and distributing elements using point clouds as the medium.

Figure 7.4: Point cloud of the rockery at Rosedown, Louisiana

7.2.1 Library

This book uses a library of laser-scanned plants to model landscapes. In computer graphics, 3-dimensional models of plants are used to render virtual landscapes. Plants can be procedurally generated and modeled as meshes for photorealistic rendering [53, 51, 43]. There are massive libraries of procedurally generated trees such as Xfrog [23, 52, 49], SpeedTree [10], and Laubwerk [4]. While software for procedurally generating plants and libraries of procedurally generated plants are used extensively in digital cinema, computer games, and design, this book uses laser-scanned plants instead.

Point clouds have a peculiar aesthetic – a blend of abstraction and realism – that makes them a powerful medium for the creative design of landscapes [9]. While procedurally generated plants are photorealistic simulations, laser-scanned point clouds are abstract representations of reality. As data, scanned point clouds are direct, indexical, measured records of the physical world. As representations, however, they are abstract, like pointillist paintings, in their incompleteness and discontinuity. When point clouds are used as a design medium, they simultaneously evoke a sense of both realism and abstraction. While a realistic rendering connotes a fixed, finished design, an abstract rendering suggests an incomplete work in progress. This sense of unfinished contingency encourages continued design exploration and evokes the dynamic, ever changing character of landscape.

A free, open source library of point clouds has been compiled as a companion to this book. This library – Cloud Forest – is a collection of laser-scanned plants. Each plant is available in `.xyz`, `.e57`, `.laz`, `.pcd`, and `.3dm` formats. Download the library from https://doi.org/10.5281/zenodo.8194066 or https://xyz.cct.lsu.edu.

Laurus nobilis
Bay Laurel

Magnolia soulangeana
Saucer Magnolia

Magnolia soulangeana
Saucer Magnolia

Morella californica
Wax Myrtle

Polystichum acrostichoides
Christmas Fern

Quercus alba
White Oak

Figure 7.5: Cloud Forest library

7.2.2 Modeling with Grasshopper

Model a simple computational planting design with Grasshopper. In Rhino 8, Grasshopper has native support for point clouds. For older versions of Rhino, there are Grasshopper plugins for point cloud processing including `Volvox` [34, 36], `Tarsier` [27], and `Cockroach` [18]. Import plants from the point cloud library into Rhino and transform them with Grasshopper to generate a simple computational planting design. Use the Grasshopper definition `point-cloud-planting.gh` as a guide for randomly scattering the ferns beneath the tree.

Data: `laurus-nobilis-01-s.3dm` & `polystichum-acrostichoides-01-s.3dm`

Code: `point-cloud-planting.gh`

Import In Rhino, import a tree and a fern from the point cloud library. First start a new Rhino document with the units set to meters. In Rhino's command line run the `Import` command to import `laurus-nobilis-01-s.3dm` into the document on a new layer. Then import `polystichum-acrostichoides-01-s.3dm` onto another layer. Rhino can also import point clouds in `.xyz` or `.e57` format.

```
Import
```

Start Rhino's visual programming interface with the `Grasshopper` command. The point clouds will be stored in Grasshopper parameters. Grasshopper for Rhino 8 has a new `Point Cloud` parameter. For older versions of Rhino, install the plugin `Volvox` and use its `Cloud` parameter instead. Add `Point Cloud` parameters to the canvas. Right click on each parameter and select a point cloud from the Rhino scene.

Tree

Fern

Transformation Use Grasshopper to randomly scatter ferns at the foot of the tree. First create a random population of points in a circular region at the base of the tree. Then move a copy of the fern to each of the random points. Finally randomly scale and rotate the scattered ferns to give them some variety. Bake the resulting point clouds to create geometry in Rhino. Move copies of the fern to random locations in a circular region. First, create a circle with the `Circle` component. Then plug the circle into a `Boundary Surfaces` component to create a planar circular region. Generate a random set of points inside this region with the `Populate Geometry` component. Set the input geometry to the circular surface and set the count and seed with number sliders. These points will be used to define translation vectors for moving copies of the fern. Plug the `Point Cloud` parameter storing a fern into a `Move` component. Set this component's input motion to the random population of points. This will define translation vectors from the origin at `0,0,0` to each of the random points.

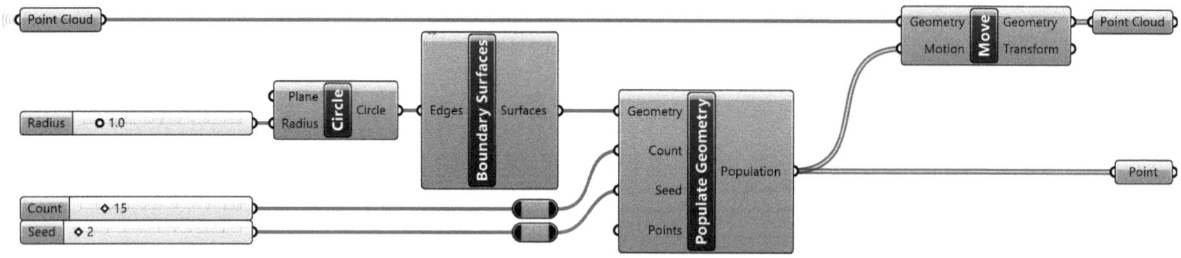

Randomly scale the scattered ferns. Plug the translated point clouds into a `Scale` component as the input geometry. Set the center of scaling to the random population of points. Set the scale factor with a list of random values generated by a `Random` component. Set the random range from `0.5 To 1.0` with a text panel, set the number of random values to match the number of points, and optionally set a seed for the random number generator.

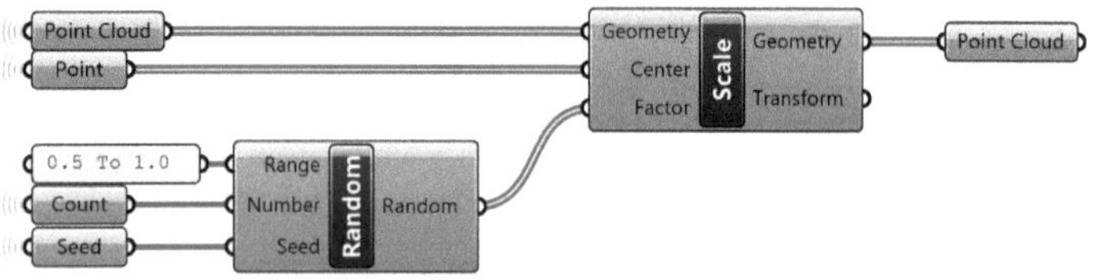

Randomly rotate the scattered, scaled ferns. Plug the scaled point clouds into a `Rotate` component. Set the rotation plane to the random population of points. Set the rotation angle to a list of random values from `0 To 360`. Right click on the rotation angle to set its units to degrees. Right click to bake the resulting geometry. This will generate a set of point clouds in Rhino for visualization and export.

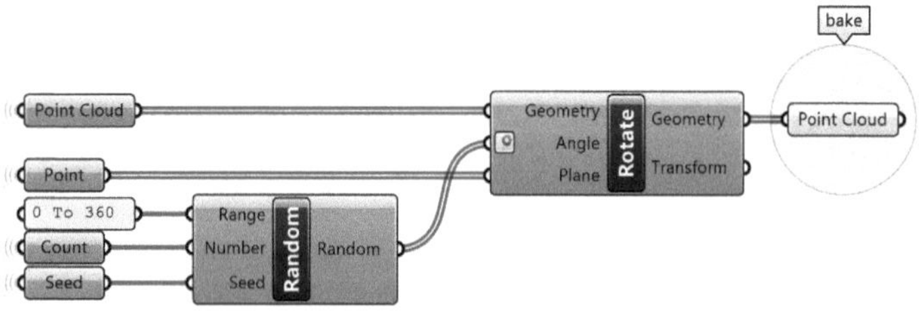

7.2.3 Visualization

While Rhino can be used for simple point cloud visualizations, other software will have more options for rendering point clouds. This book uses free and open source software for point cloud processing and visualization. CloudCompare is used [3, 47, 45] to edit point clouds and Potree [15] to visualize point clouds. Other software with robust support for visualizing point clouds include Veesus Arena4D and its plugin for Rhino [7], Krakatoa [33], and Unreal Engine with the LiDAR Point Cloud plugin [20].

Rhino In Rhino use the command `ViewCaptureToFile` to save a rendering of a point cloud as a `.png` image file. Set a transparent background and a scale factor.

Potree Potree is a viewer for rendering massive point clouds [15, 35, 21]. It uses an octree data structure to handle massive point clouds efficiently. Point clouds are rendered as high quality splats with adaptive point sizing and eye-dome lighting. Rather than rendering points as individual pixels or circles, clusters of points are blended together into splats to represent surfaces [48, 46]. While Potree is typically used for interactive web visualizations of point clouds, there is also a desktop version. To use PotreeDesktop [16], drag and drop a `.laz` point cloud into the viewer. Since Rhino can only export point clouds in `.xyz` format, use Python, CloudCompare, or a Grasshopper plugin such as Volvox to convert them to `.laz` format.

7.2.4 Modeling with Rhino Scripting

Python can be used to scatter plants with more nuances. Use Python scripting in Rhino to create a Gaussian distribution of ferns from the point cloud library. Download the model of a fern `polystichum-acrostichoides-01-s.3dm` from the library. Import this model into a new document in Rhino with units set to meters. Open Rhino's `ScriptEditor` and start a new Python script.

```
ScriptEditor
```

Data: `polystichum-acrostichoides-01-s.3dm`

Code: `point-cloud-planting.py`

Libraries At the beginning of the script, import Numpy and RhinoScriptSyntax.

Gaussian Distribution Generate a set of 2-dimensional points with a Gaussian or normal distribution. First instantiate Numpy's random number generator. Then generate normal distributions of `n` values with means of zero and standard deviations of 0.5 using the function `numpy.random.Generator.normal`. Assign a normal distribution to `x` and another to `y`. Generate an array of zeros with the function `numpy.zeros` and assign that to `z`. Interweave the arrays of x-, y-, and z-coordinates together with the function `numpy.column_stack`. Create points from these coordinates using the function `rs.AddPoints`

Selection Select the point cloud of the fern `polystichum-acrostichoides-01-s` in your Python script. Prompt the user to manually select a point cloud from the scene with the function `rs.GetObjects`. Use the parameter `rs.filter.pointcloud` to allow only point clouds.

```
obj = rs.GetObject("Select point cloud", rs.filter.pointcloud, preselect=True)
```

Layer Alternatively, select the point cloud automatically by its layer in the Rhino scene. Use the function `rs.ObjectsByLayer` to automatically select all objects on a given layer. Make sure that the point cloud is on the correct layer, that there is no other geometry on that layer, and that layer names in the script and scene match. Then use the function `rs.GetObjects` with the parameter `rs.filter.pointcloud` to return a point cloud object from the layer.

```
layer = "polystichum-acrostichoides-01-s"
obj = rs.ObjectsByLayer(layer, True)
obj = rs.GetObject("Select point cloud", rs.filter.pointcloud, preselect=True)
```

Transformations Move, rotate, and scale copies of the selected point cloud. Loop through the list of scattered points and move a copy of the fern to each point. Use the function `rs.CopyObject` to create a copy of an object at a given point. Use the function `rs.RotateObject` to rotate the copy around a given point. Generate a random value between 0 and 360 for the rotation angle using the function `numpy.random.Generator.uniform`. Finally use the function `rs.ScaleObject` to scale the copy. Set the center of scaling to the point and generate a scale factor from a Gaussian distribution with a mean of 0.75 and a standard deviation of 0.25. Take its absolute value with `numpy.absolute` to avoid generating a negative scale factor.

```python
# requirements: numpy
import numpy as np
import rhinoscriptsyntax as rs

# instantiate random number generator
rng = np.random.default_rng()

# generate x-, y-, and z-values
n = 10
mu = 0 # mean
sigma = 1.5 # standard deviation
x = rng.normal(mu, sigma, n)
y = rng.normal(mu, sigma, n)
z = np.zeros(n)

# interweave coordinates
coordinates = np.column_stack((x, y, z))

# create points
points = rs.AddPoints(coordinates)

# select objects by layer
layer = "polystichum-acrostichoides-01-s"
obj = rs.ObjectsByLayer(layer, True)
obj = rs.GetObjects(
    "Select point cloud",
    rs.filter.pointcloud,
    preselect=True
    )

# loop
mu = 0.75
sigma = 0.25
for point in points:

    # set variables
    angle = rng.uniform(0.0, 360.0)
    factor = np.absolute(rng.normal(mu, sigma))
    scale = (factor, factor, factor)

    # transform
    transform = rs.CopyObject(obj, point)
    transform = rs.RotateObject(transform, point, angle)
    transform = rs.ScaleObject(transform, point, scale)
```

7.2.5 Modeling with Grasshopper Scripting

Use Python scripting in Grasshopper to randomly scatter a grove of trees with a Gaussian distribution.

Data: `laurus-nobilis-01-s.3dm` & `polystichum-acrostichoides-01-s.3dm`

Code: `gaussian-planting.gh` & `gaussian-planting.py`

Inputs Begin by importing a tree and a fern from the point cloud library into Rhino. Then in Grasshopper create two `Point Cloud` parameters. Set the first parameter to the tree and the second to the fern.

Scatter Trees Create a script to scatter the trees. Add a `Script` component to the canvas. For inputs use `Boolean Toggle` to run the script, set the input point cloud to the tree, set an input point to scatter around, set the number of plants to scatter, and set the standard deviation for the Gaussian distribution. In Grasshopper, scripts expect their parameters to be single items. List or data trees must be specified explicitly by setting type hints. In this example, a list of points will be passed between scripts, so right click on the input `Points` parameter and set its type hint to list. The outputs of the script will be a list of points and point clouds. Double click on the `Script` component to start writing the code in the `ScriptEditor`.

After importing the libraries and setting the script context, use an **`if`** statement to run the script when the `Boolean Toggle` is activated. Create an empty 3D array with `numpy.empty`. Then loop through the list of input points and create random normal distributions centered around each point for the x- and y-coordinates, while setting the z-coordinates to zero. Stack the coordinates into a 3D array and then append them to the empty array with `numpy.append`. Once the **`for`** loop finishes, create points from this array of coordinates using the function `rs.AddPoints`. Then create an empty list of point clouds and loop through the points again. Copy an instance of the input point cloud to each point, randomly rotate that point cloud around that point, and then randomly scale the point cloud. Append the point clouds to the list. Save and close the script.

Scatter Ferns Now create a Gaussian distribution of ferns for the understory. Copy the script, set the input point cloud to the fern, and set the input points to the output points of the previous script. This will scatter ferns around the base of each tree. Increase the number of plants to scatter and decrease the standard deviation for a tighter scattering.

```
# requirements: numpy
import numpy as np
import Rhino
import rhinoscriptsyntax as rs
import scriptcontext as sc
sc.doc = Rhino.RhinoDoc.ActiveDoc

if Run:

    # create empty array
    coordinates = np.empty(shape=[0, 3])

    # loop
    for point in Points:

        # instantiate random number generator
        rng = np.random.default_rng()

        # generate random x-, y-, and z-values
        n = int(Count)
        sigma = float(Deviation) # standard deviation
        x = rng.normal(point[0], sigma, n)
        y = rng.normal(point[1], sigma, n)
        z = np.zeros(n)
        xyz = np.column_stack((x, y, z))
```

```
    # append arrays
    coordinates = np.append(coordinates, xyz, axis=0)

# create points
Points = rs.AddPoints(coordinates)

# loop
Clouds = []
for point in Points:

    # set variables
    angle = rng.uniform(0.0, 360.0)
    s = rng.uniform(0.5, 1.0)
    scaling = (s, s, s)

    # transform
    transform = rs.CopyObject(Cloud, point)
    transform = rs.RotateObject(transform, point, angle)
    transform = rs.ScaleObject(transform, point, scaling)
    Clouds.append(transform)
```

7.3 Programming

With scripting designers can go beyond the limits of design software and develop new design processes. If a feature is missing, they can develop it. For example, while Rhino does not support point cloud formats like `.las` and `.laz`, this data can easily be imported with Python scripting. This section covers importing and visualizing `.laz` point clouds with Python scripting.

7.3.1 Lidar with Python

As a high-level programming language with extensive libraries for data science, Python can be used to easily read, write, and visualize point clouds. Python libraries for reading and writing point clouds include Laspy [2] and Pyntcloud [12], while libraries for visualizing point clouds include PyVista [28, 6], Open3D [31], and Mayavi [41]. In this section, Laspy is used with its optional dependency `laz-rs` for decompressing `.laz` files and PyVista.

Reading Use the Laspy library to read `.las` or `.laz` point clouds as Numpy arrays. First import Laspy. Then set the path to a `.laz` file. See appendix C for a guide to properly specifying file paths. Use the function `laspy.read` to read the `.laz` file. This will return a `laspy.LasData` object. Use the `laspy.lasdata.LasData.xyz` method to return a Numpy array of x-, y-, and z-coordinates.

Data: `laurus-nobilis-01-l.laz`

Code: `lidar-read.py`

```
# import modules
import laspy

# set path
data = 'laurus-nobilis-01-l.laz'

# read lidar
las = laspy.read(data)
print(las.xyz)
```

```
[[-0.13818385 -0.16506511 -0.02802961]
 [-0.13587119 -0.16573959 -0.02693765]
 [-0.12941553 -0.16782719 -0.02658384]
 [-0.14753009 -0.15494781 -0.02590959]
 [-0.14502479 -0.15013033 -0.02684133]
 [-0.1400788  -0.1586414  -0.02725809]
 [-0.13943626 -0.15796691 -0.02610223]
 [-0.13291646 -0.16105038 -0.02513902]
 [-0.12851621 -0.16304165 -0.02696913]
 ...]
```

Rendering Render the point cloud using PyVista. First import Laspy and PyVista, set the path to a `.laz` file and read the data. Optionally, set a theme for plotting with `pyvista.set_plot_theme`. Use the `document` theme with a white background. Plot the coordinates using the function `pyvista.plot`. Set the first parameter – the object to be plotted – to the Numpy array of coordinates. Optionally set other parameters such as point size, window size, camera position, and eye-dome lighting.

Data: `laurus-nobilis-01-l.laz`

Code: `lidar-render.py`

```
# import modules
import laspy
import pyvista as pv

# set path
data = 'laurus-nobilis-01-l.laz'

# read lidar
las = laspy.read(data)
xyz = las.xyz

# set plot theme
pv.set_plot_theme('document')

# plot with eye-dome lighting
pv.plot(
    xyz,
    point_size=10,
    color='white',
    window_size=(4000, 4000),
    cpos='xz',
    show_axes=False,
    eye_dome_lighting=True
    )
```

Color Rendering Each dimension of the point cloud including color channels is stored as attributes of the `laspy.LasData` object. Simply call `las.dimension_name` to return a dimension. If the point cloud has colors, they can be accessed by calling `las.red`, `las.green`, and `las.blue`.

Data: `magnolia-soulangeana-02-l.laz`

Code: `lidar-render-rgb.py`

Render the point cloud in color using PyVista. First import Laspy and PyVista, set the path to a `.laz` file and read the data. Return the x, y, z, red, green, and blue dimensions of the point cloud. In this example the colors need to be converted from 16-bit to 8-bit for rendering. Divide each color by 256 and recast the arrays as 8-bit unsigned integers with `numpy.ndarray.astype`. Interweave the converted colors with the function `numpy.column_stack`. Plot a colored point cloud with `pyvista.plot`. Set the object to be plotted to the array of x-, y-, and z-coordinates, set the scalars to the array of colors, and enable the `rgba` parameter to render the scalars as color.

```
# import modules
import numpy as np
import laspy
import pyvista as pv

# set path
data = 'magnolia-soulangeana-02-l.laz'

# read lidar
las = laspy.read(data)
xyz = las.xyz

# convert colors
red = (las.red/256).astype('uint8')
green = (las.green/256).astype('uint8')
blue = (las.blue/256).astype('uint8')

# plot with color
pv.set_plot_theme('document')
rgb = np.column_stack((red, green, blue))
pv.plot(
    xyz,
    scalars=rgb,
    point_size=5,
    window_size=(4000, 4000),
    cpos='xz',
    show_axes=False,
    rgba=True
    )
```

7.3.2 Lidar with Rhino Scripting

Point clouds in `.las` format can be imported directly into Rhino using Laspy and Numpy. First import Laspy, Numpy, and RhinoScriptSyntax. Then read the lidar point cloud with Laspy. If necessary convert the point cloud colors to 8-bit color. Interweave the colors with `numpy.column_stack` and then convert the array to a list. Create a point cloud in Rhino with `rs.AddPointCloud`. Set the first parameter to the array of x-, y-, and z-coordinates and set the second parameter to the list of red, green, and blue colors. This script will take time, so be patient.

Data: morella-californica-01-l.laz

Code: lidar-in-rhino.py

```
# requirements: lazrs, laspy, numpy

# import modules
import laspy
import numpy as np
import rhinoscriptsyntax as rs

# set path
data = r'morella-californica-01-l.laz'

# read lidar
las = laspy.read(data)

# convert colors
red = (las.red/256).astype('uint8')
green = (las.green/256).astype('uint8')
blue = (las.blue/256).astype('uint8')

# interweave colors
rgb = np.column_stack((
    red, green, blue
    )).tolist()

# generate colored point cloud
rs.AddPointCloud(las.xyz, rgb)
```

References

[1] [Software] Agisoft, *Metashape* version 2.0.2, 2023. url: https://www.agisoft.com/.

[2] [Software] laspy Development Team, *laspy* version 2.4.1, 2023. lic: laspy license. url: https://laspy.readthedocs.io/en/latest, vcs: https://github.com/laspy/laspy.

[3] [Software] Daniel Girardeau-Montaut, *CloudCompare* version 2.13, 2023. lic: GNU General Public License v2. url: https://www.danielgm.net/cc/, vcs: https://github.com/CloudCompare/CloudCompare.

[4] [Software] Laubwerk, *Laubwerk Plant Kits* version 1.0.45, 2023. url: https://www.laubwerk.com/.

[5] [Software] OpenDroneMap, *OpenDroneMap* version 3.1.0, 2023. lic: GNU Affero General Public License v3.0. url: https://opendronemap.org/, vcs: https://github.com/OpenDroneMap/ODM.

[6] [Software] PyVista Development Team, *PyVista* version 0.40.0, 2023. lic: MIT License. doi: 10.5281/zenodo.8101943, url: https://pyvista.org/, vcs: https://github.com/pyvista/pyvista.

[7] [Software] Veesus, *Veesus Point Clouds for Rhino* version 2023.1.8501.19194, 2023. url: https://veesus.com/.

[8] FARO. *Faro Focus Premium*. 2022. url: https://www.faro.com (visited on 01/18/2023).

[9] Brendan Harmon and Nicholas Serrano. "Point Cloud Aesthetics". In: *Journal of Digital Landscape Architecture* 7 (2022), pages 335–344. issn: 2367-4253. doi: 10.14627/537724033.

[10] [Software] Interactive Data Visualization, *SpeedTree* version 9.0.1, 2022. url: https://store.speedtree.com/.

[11] [Software] Pix4D, *Pix4D* version 4.8.2, 2022. url: https://www.pix4d.com/.

[12] [Software] Pyntcloud Developers, *Pyntcloud* version 0.3.1, 2022. lic: MIT License. url: http://pyntcloud.readthedocs.io/, vcs: https://github.com/daavoo/pyntcloud.

[13] [Software] Rapidlasso, *LASzip* version 2.0.2, 2022. url: https://rapidlasso.de/laszip/, vcs: https://github.com/LAStools/LAStools.

[14] RIEGL Laser Measurement Systems. *RIEGL VZ-2000i Datasheet*. 2022. url: http://www.riegl.com (visited on 01/18/2023).

[15] [Software] Markus Schütz, *Potree: WebGL point cloud viewer for large datasets* version 1.8.1, 2022. lic: BSD 2-Clause License. vcs: https://github.com/potree/potree/.

[16] [Software] Markus Schütz, *PotreeDesktop* version 1.8.1, 2022. lic: BSD 2-Clause License. vcs: https://github.com/potree/PotreeDesktop/.

[17] Philipp R.W. Urech, Aurel von Richthofen, and Christophe Girot. "Grounding landscape design in high-resolution laser-scanned topography". In: *Journal of Landscape Architecture* 17.1 (2022), pages 58–69. issn: 1862-6033. doi: 10.1080/18626033.2022.2110422.

[18] [Software] Petras Vestartas and Andrea Settimi, *Cockroach: A plug-in for Rhino point cloud post-processing and meshing* version 1.0.0, 2021. EPFL Laboratory for Timber Constructions. lic: GNU Lesser General Public License v3.0. url: https://www.food4rhino.com/en/app/cockroach, vcs: https://github.com/9and3/Cockroach.

[19] Tim D'Urban Jackson, Gareth Williams, Guy Walker-Springett, and Andrew J. Davies. "Three-dimensional digital mapping of ecosystems: a new era in spatial ecology". In: *Proceedings of the Royal Society B: Biological Sciences* 287.1920 (2020), page 20192383. doi: 10.1098/rspb.2019.2383.

[20] [Software Module] Epic Games, "LiDAR Point Cloud Plugin", part of *Unreal Engine* version 5, 2020. url: https://docs.unrealengine.com/5.0/en-US/lidar-point-cloud-plugin-for-unreal-engine/.

[21] Markus Schütz, Stefan Ohrhallinger, and Michael Wimmer. "Fast Out-of-Core Octree Generation for Massive Point Clouds". In: *Computer Graphics Forum* 39.7 (Nov. 2020), pages 1–13. issn: 1467-8659. doi: 10.1111/cgf.14134.

[22] Philipp R.W. Urech, Maria Angela Dissegna, Christophe Girot, and Adrienne Grêt-Regamey. "Point cloud modeling as a bridge between landscape design and planning". In: *Landscape and Urban Planning* 203 (2020), page 103903. issn: 01692046. doi: 10.1016/j.landurbplan.2020.103903.

[23] [Software] Xfrog, *Xfrog* version 6.0, 2020. url: https://www.xfrog.net/.

[24] American Society for Photogrammetry and Remote Sensing. *LAS Specification 1.4 – R15*. Technical report. Bethesda, Maryland: ASPRS, July 2019. url: https://www.asprs.org/wp-content/uploads/2019/07/LAS_1_4_r15.pdf (visited on 01/03/2023).

[25] Christophe Girot. "Cloudism". In: *Routledge Research Companion to Landscape Architecture*. Edited by Ellen Braae and Henriette Steiner. Routledge, 2019. doi: 10.4324/9781315613116.

[26] Martin Isenburg. *LASzip 3.4.3 documentation*. 2019. url: https://laszip.org/ (visited on 01/03/2023).

[27] [Software] Cameron Newnham, *Tarsier* version 1.0.1.0, 2019. lic: GNU General Public License. url: https://www.food4rhino.com/en/app/tarsier, vcs: https://bitbucket.org/camnewnham/tarsier.

[28] Bane Sullivan and Alexander Kaszynski. "PyVista: 3D plotting and mesh analysis through a streamlined interface for the Visualization Toolkit (VTK)". In: *Journal of Open Source Software* 4.37 (May 2019), page 1450. doi: 10.21105/joss.01450.

[29] Philipp R.W. Urech. "Point-Cloud Modeling: Exploring a Site-Specific Approach for Landscape Design". In: *Journal of Digital Landscape Architecture* 4 (2019), pages 290–297. issn: 2367-4253. doi: 10.14627/537663031.

[30] Felix Morsdorf, Daniel Kükenbrink, Fabian D. Schneider, M. Abegg, and Michael E. Schaepman. "Close-range laser scanning in forests: towards physically based semantics across scales". In: *Interface Focus* 8.2 (2018), page 20170046. doi: 10.1098/rsfs.2017.0046.

[31] Qian-Yi Zhou, Jaesik Park, and Vladlen Koltun. "Open3D: A Modern Library for 3D Data Processing". In: *arXiv:1801.09847* (2018). doi: 10.48550/arXiv.1801.09847.

[32] American Society for Photogrammetry and Remote Sensing. *ASPRS LAS Specification*. 2017. url: https://github.com/ASPRSorg/LAS (visited on 01/03/2023).

[33] [Software] AWS Thinkbox, *Krakatoa* version 2.12, 2017. url: https://www.awsthinkbox.com/krakatoa.

[34] [Software] Henrik Leander Evers and Mateusz Zwierzycki, *Volvox* version 0.3.0.0, 2016. Centre for Information Technology and Architecture. url: https://www.food4rhino.com/en/app/volvox, vcs: https://github.com/CITA-cph/Volvox/tree/master.

[35] Markus Schütz. "Potree: Rendering Large Point Clouds in Web Browsers". Master's thesis. Favoritenstrasse 9-11/E193-02, A-1040 Vienna, Austria: Institute of Computer Graphics and Algorithms, Vienna University of Technology, Sept. 2016. url: https://www.cg.tuwien.ac.at/research/publications/2016/SCHUETZ-2016-POT/.

[36] Mateusz Zwierzycki, Henrik Leander Evers, and Martin Tamke. "Parametric architectural design with point-clouds – Volvox". In: *Proceedings of the 34th eCAADe Conference*. Volume 2.

University of Oulu, Oulu, Finland, 2016, pages 673–682. doi: 10.52842/conf.ecaade.2016.2.673.

[37] Lewis Graham. “The LAS 1.4 Specification”. In: *Photogrammetric Engineering & Remote Sensing* 78.2 (Feb. 2012), pages 93–102. url: https://www.asprs.org/wp-content/uploads/2010/12/LAS_Specification.pdf (visited on 01/03/2023).

[38] Sameer Agarwal, Yasutaka Furukawa, Noah Snavely, Ian Simon, Brian Curless, Steven M. Seitz, and Richard Szeliski. “Building Rome in a Day”. In: *Communications of the ACM* 54.10 (Oct. 2011), pages 105–112. issn: 0001-0782. doi: 10.1145/2001269.2001293.

[39] Mathieu Dassot, Thiéry Constant, and Meriem Fournier. “The use of terrestrial LiDAR technology in forest science: application fields, benefits and challenges”. In: *Annals of Forest Science* 68.5 (2011), pages 959–974. doi: 10.1007/s13595-011-0102-2. url: https://hal.archives-ouvertes.fr/hal-00930668.

[40] Daniel Huber. “The ASTM E57 File Format for 3D Imaging Data Exchange”. In: *Proceedings of SPIE Electronics Imaging Science and Technology Conference (IS&T), 3D Imaging Metrology*. Volume 7864. Jan. 2011.

[41] Prabhu Ramachandran and Gaël Varoquaux. “Mayavi: 3D Visualization of Scientific Data”. In: *Computing in Science & Engineering* 13.2 (2011), pages 40–51. issn: 1521-9615.

[42] Sameer Agarwal, Yasutaka Furukawa, Noah Snavely, Brian Curless, Steven M. Seitz, and Richard Szeliski. “Reconstructing Rome”. In: *Computer* 43.6 (2010), pages 40–47. doi: 10.1109/MC.2010.175.

[43] Oliver Deussen and Bernd Lintermann. *Digital Design of Nature: Computer Generated Plants and Organics*. Springer, 2010. isbn: 978-3642073632. doi: 10.1007/b138606.

[44] Duncan P. Robertson and Roberto Cipolla. “Practical Image Processing and Computer Vision”. In: *Structure from Motion*. Edited by Margaret Varga. Wiley, 2009. Chapter 13. isbn: 9780470868546.

[45] Daniel Girardeau-Montaut. “Détection de changement sur des données géométriques tridimensionnelles”. Theses. Télécom ParisTech, May 2006. url: https://pastel.archives-ouvertes.fr/pastel-00001745.

[46] Mario Botsch, Alexander Sorkine-Hornung, Matthias Zwicker, and Leif Kobbelt. “High-quality surface splatting on today’s GPUs”. In: *Eurographics Symposium on Point-Based Graphics*. July 2005, pages 17–141. doi: 10.1109/PBG.2005.194059.

[47] Daniel Girardeau-Montaut, Michel Roux, Raphaël Marc, and Guillaume Thibault. “Change detection on point cloud data acquired with a ground laser scanner”. In: *International Archives of Photogrammetry, Remote Sensing and Spatial Information Sciences* 36 (Jan. 2005). url: https://www.danielgm.net/phd/isprs_laserscanning_2005_dgm.pdf.

[48] Matthias Zwicker, Hanspeter Pfister, Jeroen van Baar, and Markus Gross. “Surface Splatting”. In: *Proceedings of the 28th Annual Conference on Computer Graphics and Interactive Techniques*. SIGGRAPH ’01. New York, NY, USA: Association for Computing Machinery, 2001, pages 371–378. isbn: 158113374X. doi: 10.1145/383259.383300.

[49] Bernd Lintermann and Oliver Deussen. “Interactive Modeling of Plants”. In: *IEEE Computer Graphics and Applications* 19.1 (Jan. 1999), pages 56–65. issn: 0272-1716. doi: 10.1109/38.736469.

[50] David G. Lowe. “Object recognition from local scale-invariant features”. In: *Proceedings of the Seventh IEEE International Conference on Computer Vision*. Volume 2. 1999, pages 1150–1157. doi: 10.1109/ICCV.1999.790410.

[51] Oliver Deussen, Pat Hanrahan, Bernd Lintermann, Radomír Měch, Matt Pharr, and Przemyslaw Prusinkiewicz. “Realistic Modeling and Rendering of Plant Ecosystems”. In: *Proceedings*

of the 25th Annual Conference on Computer Graphics and Interactive Techniques. SIGGRAPH '98. New York, NY, USA: ACM, 1998, pages 275–286. isbn: 0-89791-999-8. doi: 10.1145/280814.280898.

[52] Oliver Deussen and Bernd Lintermann. "A modelling method and user interface for creating plants". In: *Proceedings of the Graphics Interface 1997 Conference, May 21-23, 1997, Kelowna, BC, Canada*. May 1997, pages 189–197. url: http://graphicsinterface.org/wp-content/uploads/gi1997-22.pdf.

[53] Przemyslaw Prusinkiewicz and Aristid Lindenmayer. *The Algorithmic Beauty of Plants*. Springer-Verlag, 1990. isbn: 9780387946764. url: http://algorithmicbotany.org/papers/abop/abop.pdf.

[54] Shimon Ullman and Sydney Brenner. "The interpretation of structure from motion". In: *Proceedings of the Royal Society of London. Series B. Biological Sciences* 203.1153 (1979), pages 405–426. doi: 10.1098/rspb.1979.0006.

8 Grove

8.1 Planting

Model a grove of trees that is taller and denser at its core. Start by randomly scattering trees within a boundary. Create a planting pattern that is denser at its core and sparser towards its edge by culling a growing percentage of specimens the further they are from the center of the planting area. Then scale the remaining trees by their distance from the heart of the grove.

Data: `quercus-alba-01-s.3dm` & `polystichum-acrostichoides-01-s.3dm`

Code: `random-grove.gh`

Point Clouds Import a tree and an understory plant from the point cloud library into Rhino.

8.1.1 Trees

Randomly scatter trees throughout a planting area. First create a circular region and randomly scatter points throughout the region. Then cull points that are far from the center of the region, move a point cloud tree to each point, randomly rotate each tree, and then scale the trees by their distance from the center of the region.

Region Create the boundary of the planting area with the `Circle` component. Generate a surface inside of the circle with the `Boundary Surfaces` component and then scatter random points across this surface with `Populate Geometry`. These points are potential locations for trees.

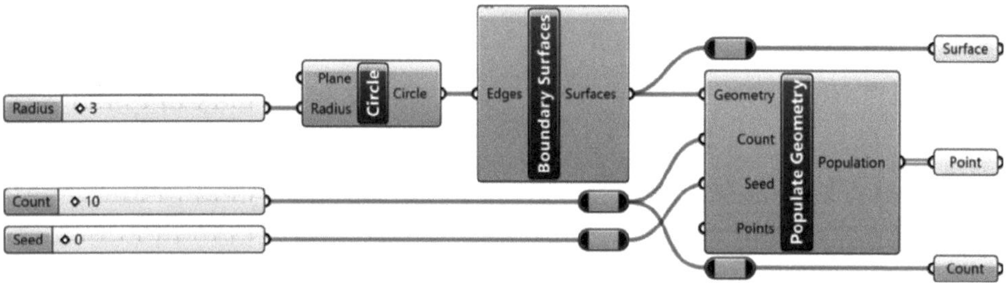

Cull some of the points that are far from the center of the circle to create a density gradient. First create a point attractor at the center of the circle. Use the `Closest Point` component to calculate the distance of each random point from the center of the circle. Set the input point to the origin `0, 0, 0`, set the input cloud to the population of random points, and set the input count to the number of random points. Plug the output closest point into a `Cull Pattern` component as the input list. For the input cull pattern, generate a list of Booleans where points that are close to the origin are more likely to be `True`, while points far from the origin are more likely to be `False`. To generate this list, plug `Closest Point`'s output distance into `Remap Numbers` as the input value. Also plug the distance into `Bounds` as the input numbers. Plug `Bounds`' output domain into `Remap Numbers` as the input source. To test this remapped list against a threshold plug it into a `Smaller Than` component as the first input. For the second input generate a list of random numbers of the same length as the population of random points using a `Random` component. Optionally trying inserting a `Graph Mapper` between the `Remap Numbers` and `Smaller Than` components to adjust the likelihood of culling distant points. Set the graph type to `Bezier` and adjust the curve to redistribute the mapped list of values.

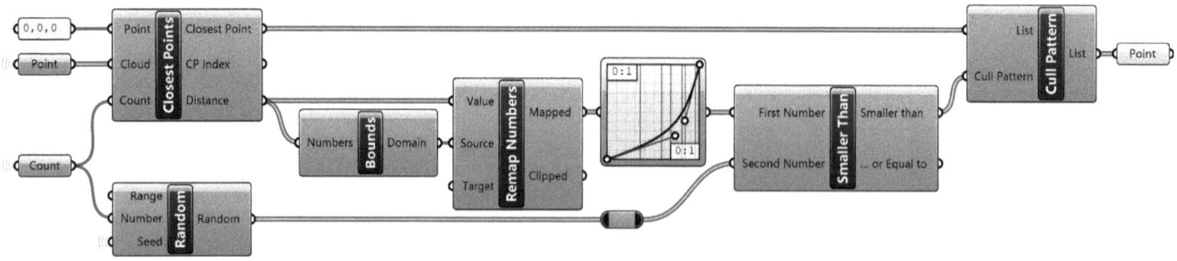

Now that the points are denser near the heart of the grove, move a point cloud tree to each point and randomly rotate the tree. First move a copy of the point cloud tree to each point with the `Move` component. Then randomly rotate each tree around its base with the `Rotate` component. Set the rotation plane to the list of culled points, so that each tree rotates around its base rather than the origin. Right click on the angle parameter and set its units to degrees. The rotation angle will be set to a list of random values between 0 and 360. To generate this list, add a `Random` component, set its range from `0 To 360`, set the number of random values based on the number of trees to rotate, and optionally set a seed for the random number generator. Calculate the number of trees by plugging the list of culled points into a `List Length` component.

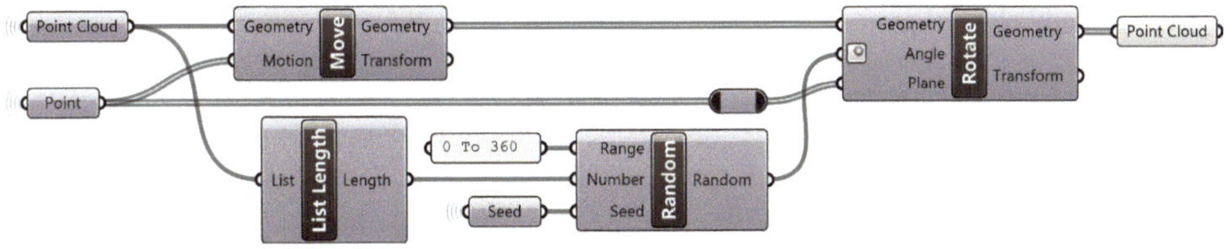

Scale Scale the trees to create a height gradient where the tallest trees are at the center of the grove, while the shortest are at the periphery. Add a `Scale` component, set the geometry to the translated and rotated list of point clouds, set the center of scaling to the list of culled points, and set the scale factor to a sequence of numbers. Use the `Range` component to generate this sequence of scaling factors. Right click on the output range to reverse the list, set a domain such as `0.75 To 1`, and set the steps based on the number of culled points. Right click on the steps parameter and set the expression `x-1`. Calculate the number of steps by plugging the list of culled points into a `List Length` component. To create a non-linear sequence, plug the output range into a `Graph Mapper` component, set the graph type to `Bezier` curve, and draw a custom curve.

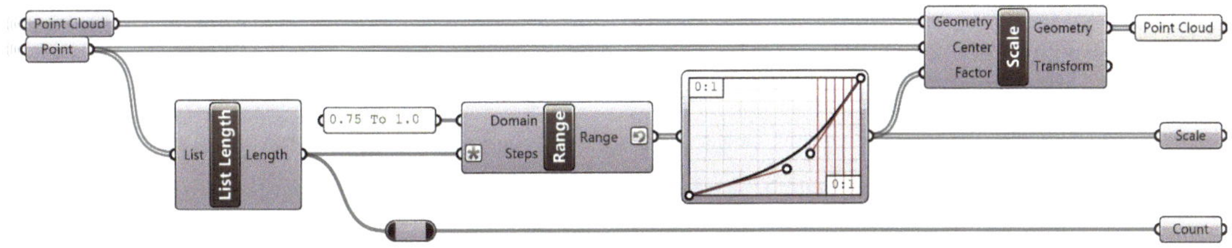

8.1.2 Understory

Randomly scatter understory plants underneath the trees.

Scatter Start by defining understory planting areas around the base of each tree. Use the `Circle` component to draw circles around the base of each tree. Set the base plane to the list of culled points and the radius to the scaling factor that was used to scale the trees. This will create understory planting areas that are proportional to the size of the trees. Scale these circles to create an inner boundary so that the understory plants are not too close to the trunks of the trees. Add a `Scale` component. Set the geometry to the list of circles, the center of scaling to the culled list of points, and the scale factor with a number slider. Then interpolate a planar surface between the inner and outer circles with the `Boundary Surfaces` component. Generate a random population of points on this surface with the `Populate Geometry` component. Understory plants will be distributed at each of these points. Set the count to a list of random numbers generated with the `Random` component. Set a domain such as `6 To 12`, set the number to the length of the culled list of points, and optionally set a seed for the random number generator.

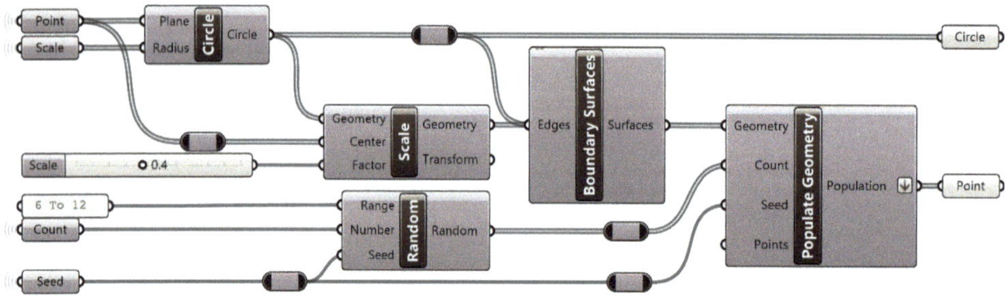

Transform Move a copy of the point cloud – a laser-scanned fern – to each point in the random population. Plug the point cloud into a `Move` component as the input geometry. Set the input motion to the list of random points. Then randomly scale and rotate the translated point clouds. Plug the translated point clouds into a `Scale` component as the input geometry. Set the center of scaling to the list of random points, and set a random scaling factor. Generate a list of scaling factors with a `Random` component. Then plug the scaled point clouds into a `Rotate` component. Set the rotation plane to the list of random points. Right click on the input angle parameter to set the units to degrees. To set the rotation angle multiply the list of random values used for scaling by `360`.

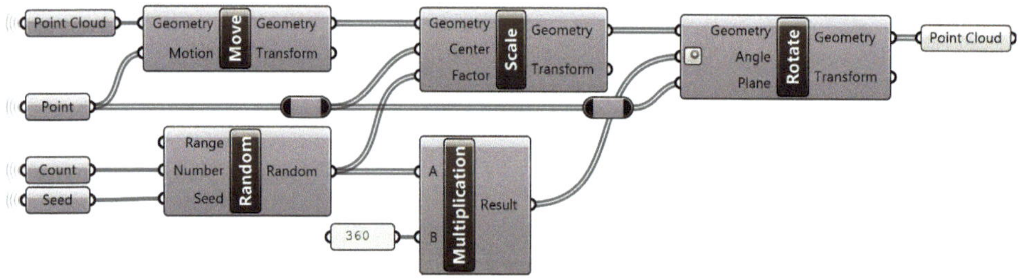

8.2 Paving

Create a paving pattern with lighter pavers near the trees and darker pavers far from the trees. After defining the area for paving, generate a grid of pavers and then remove pavers that extend beyond the paving area. Finally, assign a color gradient to the pavers based on their distance from the trees.

Region Create a circular region to demarcate the extent of the paving area with a `Circle` component. Meanwhile, merge intersecting planting areas with a `Region Union` component. Then interpolate a planar surface between the outer boundary of the paving area and the planting areas with a `Boundary Surfaces` component.

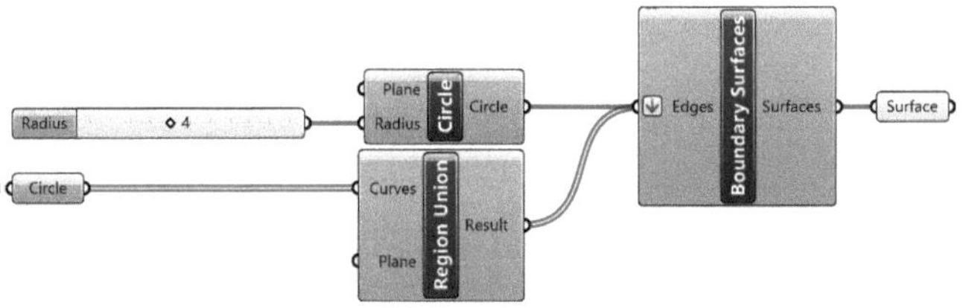

Cull Generate a grid of points with a `Square` component. Flatten the output data tree of points into a list. Then use a `Surface Closest Point` component to find the distance between the points and the boundary surface. Use a `Smaller Than` component to test whether points are within or very close to the surface; compare the distance against a threshold such as `0.01`. Cull points that greater than this threshold with a `Cull Pattern` component.

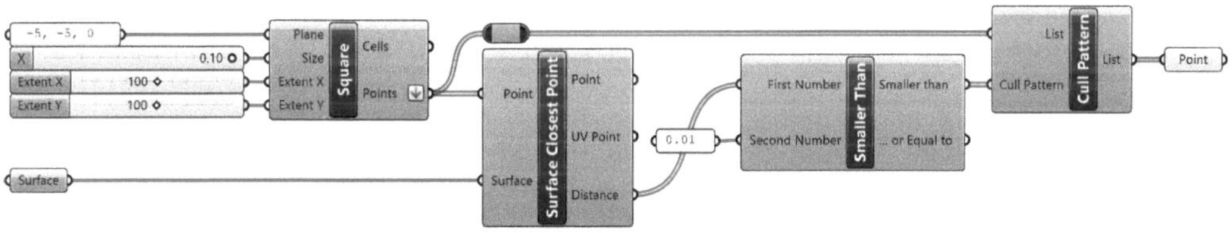

Pavers Model a paver at each remaining point in the culled grid. Use a `Center Box` component to model the pavers. Set the base of the box to the list of points and use number sliders to define the x-, y-, and z-dimensions of the box.

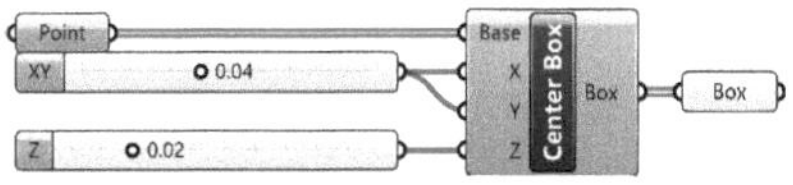

Gradient Use the trees as attractors to generate a color gradient for the paving. Create an attractor with an `Closest Point` component. Set the input point parameter to the list of culled grid points and set the cloud parameter to the points at the base of the trees. Remap the output distance to a domain from `0 To 1` with a `Remap Numbers` component. Find the source domain for remapping by plugging the distance into a `Bounds` component. This remapped list of distances will be used to assign a color gradient to the pavers. To add some variety, randomly jitter this list. Use a `Multiplication` component to multiply the remapped numbers by list of random numbers. Generate the random numbers with a `Random` component. Use a `List Length` component to calculate the number of random values based on the number of culled grid points. To increase the contrast in the paving gradient, multiply the jittered, remapped list of distances by a factor. Plug the list and a number slider into a `Multiplication` component. Then plug the result into a `Gradient`. Finally add a `Custom Preview` and set the input geometry to the list of boxes and the input material to the color gradient.

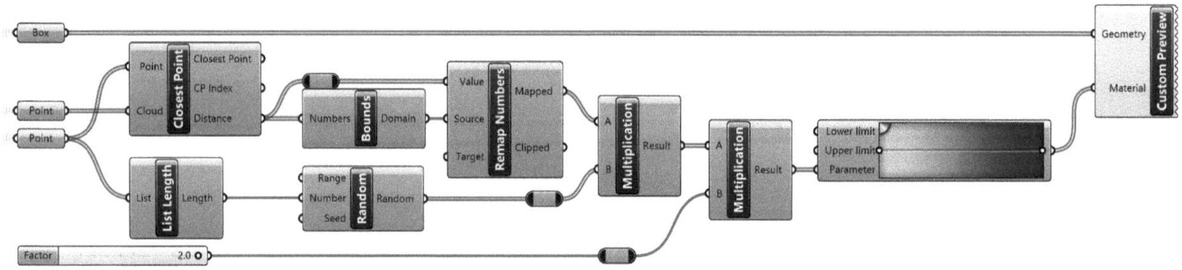

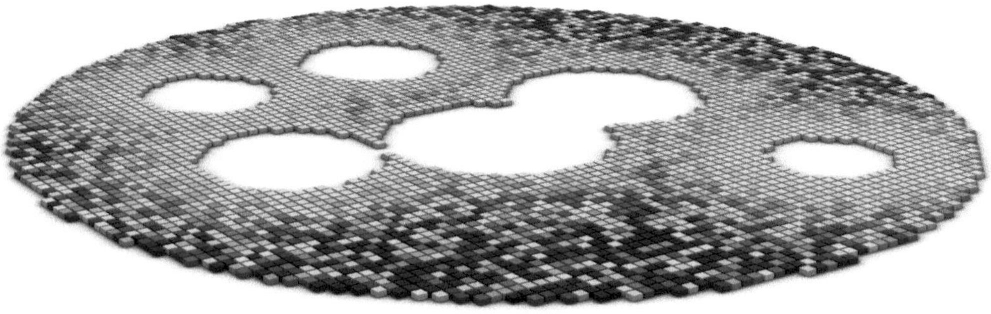

9 Waves

9.1 Waves

Mathematics are the foundation of generative art and design. Mathematical functions can be used to generate an infinite variety of shapes and forms. This chapter covers sinusoidal functions, which describe the form of waves. Waves – such as wind, water, sound, and light – are propagating processes of disturbance and recovery in which energy oscillates through time and space. Topography – the shape of the Earth's surface – is not only shaped by waves, but can also act as waves. Wind can form continuously shifting fields of barchan dunes that move as waves across the desert. As wind energy is transferred into waves of sand, propagating from grain to grain and dune to dune, the dunes migrate and coalesce as oscillating waves [10, 11].

Waves can play a generative role in design. Designers have modeled landforms based on waveforms and designed landforms to interact with waves. With Wave Field, Maya Lin designed a field of landforms – based on Stokes' mathematical model of waves [13] – to express dynamism and continuous motion [12, 9]. With the MAX IV Laboratory Landscape, Snøhetta used waves not only as form, but also as process. Their wave field of landforms around the laboratory was designed to attenuate vibrations from the nearby highway [6].

9.2 Sine Waves

Sine and cosine are used to describe oscillating, periodic, i.e. regularly repeating, phenomena such as waves. In physics, sine waves are a function of time that can be graphed by plotting time along the x-axis and the amplitude of the wave along the y-axis. While the sine function $y = \sin(x)$ has an infinite domain, its domain can be sampled between, for example, 0 and 2π. In that case the sine function can be graphed as a wave by plotting $y = \sin(x)$ for $\{x \mid 0 \leq x \leq 2\pi\}$.

$$y = \sin(x) \quad \textbf{Equation } 9.1$$

9.2.1 Sine Waves in Grasshopper

To graph a basic sine wave in Grasshopper, generate a sequence of x-values from 0 to 2π, then compute y as the sine of each x-value, construct points from the x- and y-values, and finally interpolate a curve through the points. Use the Grasshopper definition `waves.gh` as a guide.

Code: `waves.gh`

sine	**cosine**	**unit circle**
$\{x \mid 0 \leq x \leq 2\pi\}$	$\{x \mid 0 \leq x \leq 2\pi\}$	$\{t \mid 0 \leq t \leq 2\pi\}$
$y = \sin(x)$	$y = \cos(x)$	$x = \cos(t)$ $y = \sin(t)$

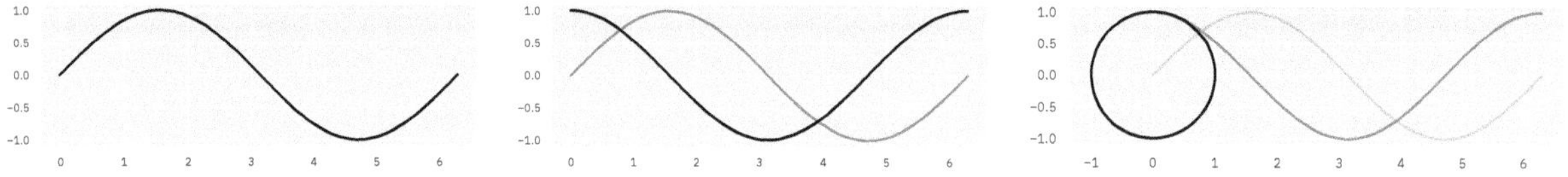

Figure 9.1: Sine waves

Sine To draw a sine wave with Grasshopper `Interpolate` a curve through a series of points defined by a `Range` of x-coordinates from 0 to 2π and the `Sine` of a range of y-coordinates from 0 to 2π. Start by placing an `Interpolate` component. Then add a construct point component and plug the output points into `Interpolate`'s input vertices. Now add a `Range` component and plug the output range into `Construct Point`'s input x-coordinate. This will generate a straight line of points along the x-axis. Set the domain of the `Range` component to 2π and the steps to 200. Finally add a `Sine` component, plug the `Range` in as its input list of values, and plug the resulting list into `Construct Point`'s input y-coordinate to generate a sine wave.

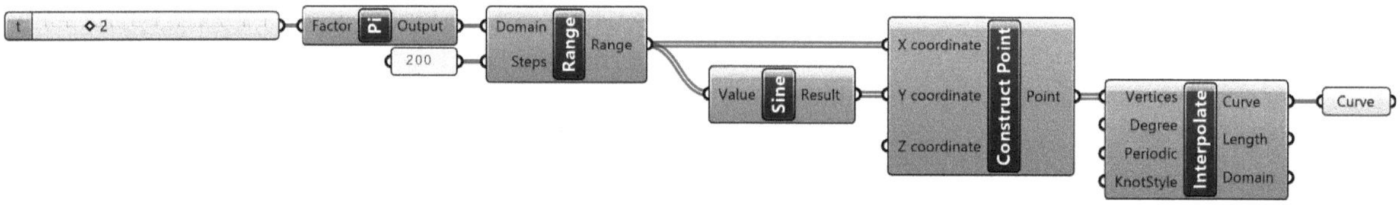

Cosine The Grasshopper definition for drawing a cosine wave is the same as the definition for drawing a sine wave, except the `Sine` component is replaced by the `Cosine` component. Interpolate a curve through a set of points defined by a range of x-coordinates from 0 to 2π and the `Cosine` of a range of y-coordinates from 0 to 2π.

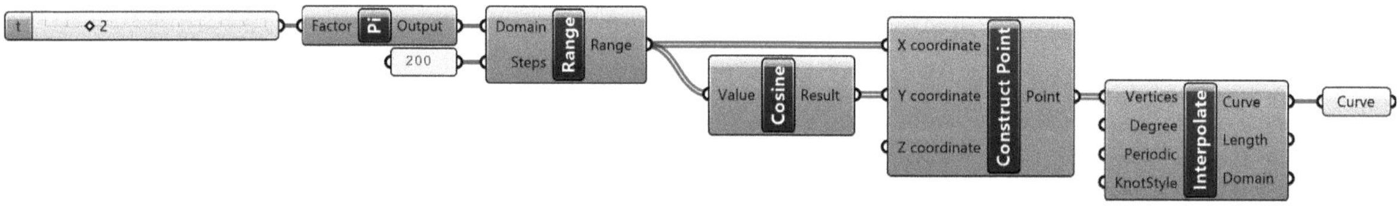

Unit Circle The unit circle can be plotted as the parametric equations $x = \cos(t)$ and $y = \sin(t)$ where t is time. To draw the unit circle with Grasshopper interpolate a curve through a series of points defined by the `Sine` of a range of x-coordinates from 0 to 2π and the `Cosine` of a range of y-coordinates from 0 to 2π. Start by composing a Grasshopper definition for a cosine wave. Then plug a `Sine` component with a range of values from 0 to 2π into `Construct Point`'s x-coordinate.

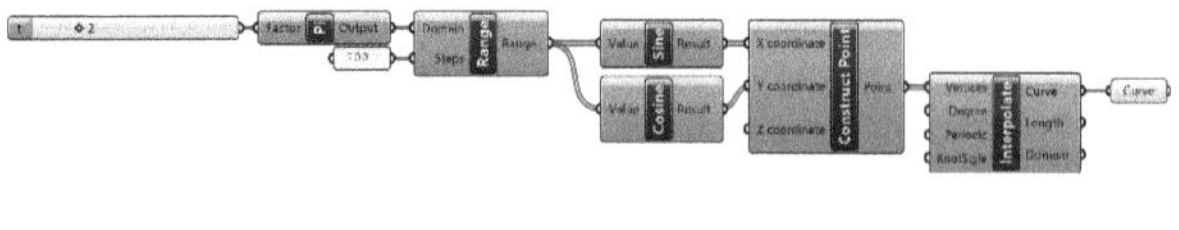

Expressions Math can be written more concisely in Grasshopper using expressions. An entire equation can be composed using a single `Expression` component from the script section of the math tab. Try drawing a sine curve using an expression in Grasshopper. Start by composing a Grasshopper definition for a sine curve. Then replace the `Sine` component with an `Expression`. Edit the input parameters by right clicking on the left side of the `Expression` component. Rename `variable x` to `x` and remove `variable y`. Then open the expression editor by double clicking on the center of the component and write the expression `sin(x)`. While this is a simple example, expressions can make Grasshopper definitions for complicated functions more concise and legible. Learn more about expressions with the Grasshopper Primer [8].

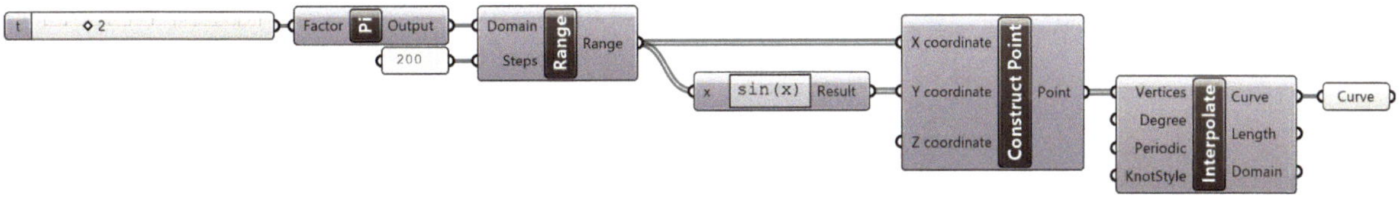

9.2.2 Sine Waves in Python

To graph a basic sine wave in Python, first import the necessary libraries, then generate a sequence of x-values, compute y as the sine of each x-value, and finally plot the x- and y-values as coordinates. Start by importing the `NumPy` library for performing mathematical operations on arrays [5, 1] and the `Seaborn` library for elegantly visualizing data [4, 3]. Set the default Seaborn theme with the function `seaborn.set_theme`. To create a domain from 0 to 6π, use the function `numpy.linspace` to generate an evenly spaced sequence as a 1-dimensional array. Set its start parameter to 0, its end parameter to 6π using `numpy.pi`, and the number of samples to 100. Assign the resulting array to the variable x. Then compute $y = \sin(x)$ using the function `numpy.sin`. Assign the resulting array to the variable y. Plot the x and y variables along the x- and y-axes with the function `seaborn.scatterplot` or `seaborn.lineplot`.

Code: `sine-wave.py`

Python

```python
# import libraries
import numpy as np
import seaborn as sns
sns.set_theme()

# plot sine wave
x = np.linspace(0, 6*np.pi, 100)
y = np.sin(x)
sns.scatterplot(x=x, y=y)
```

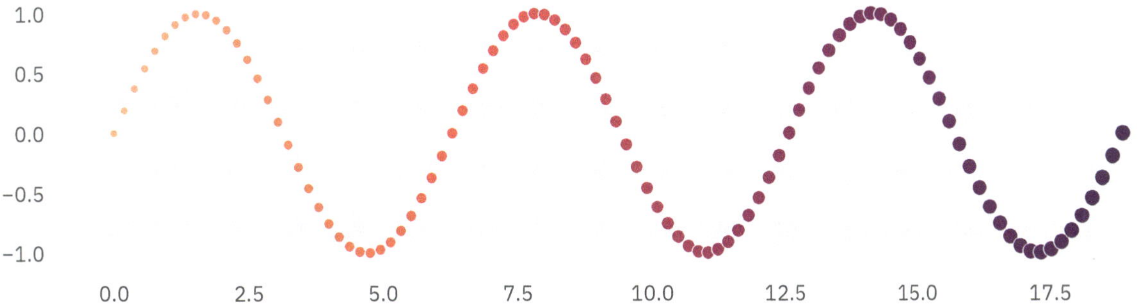

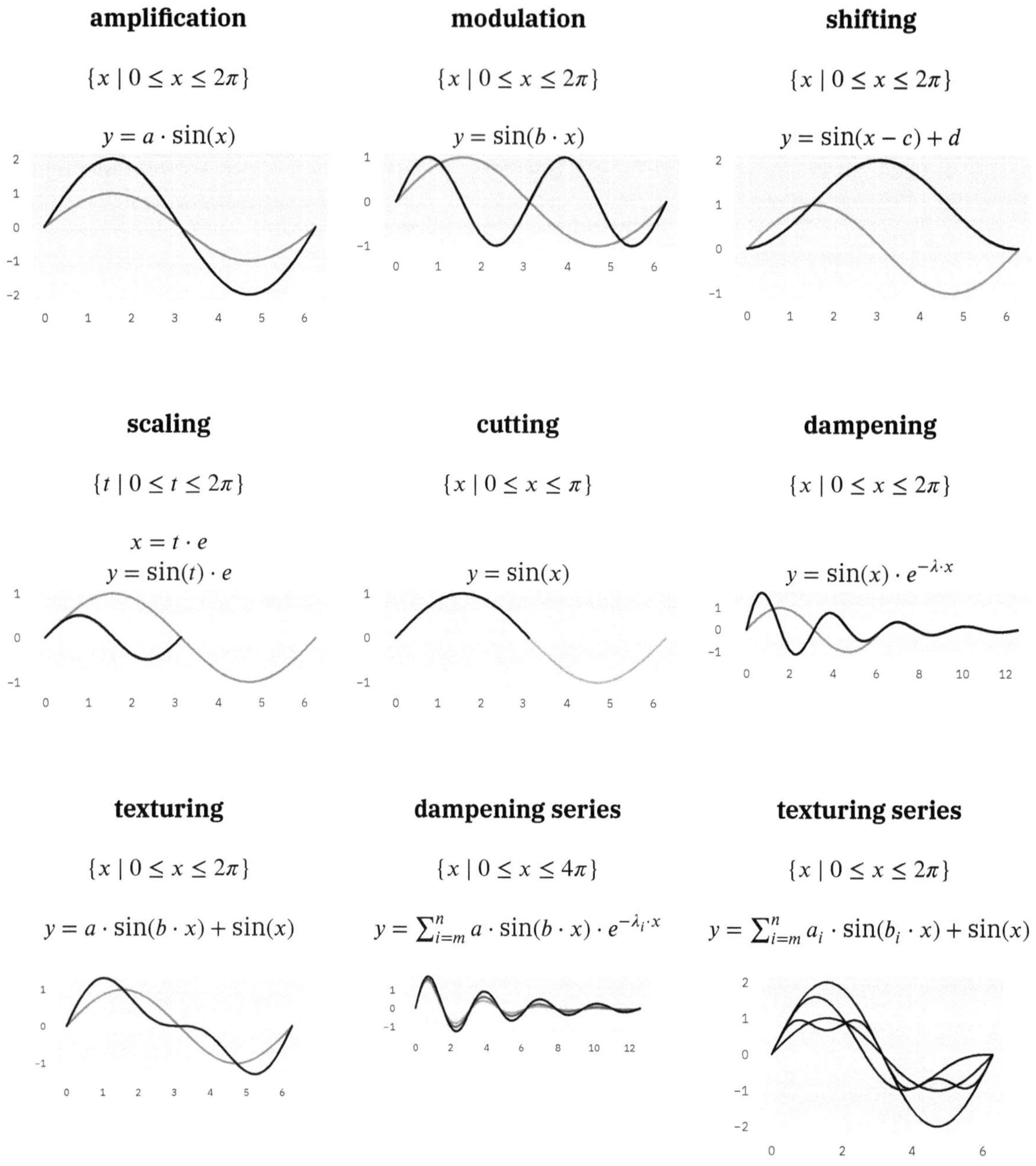

Figure 9.2: Transformations

9.3 Transformations

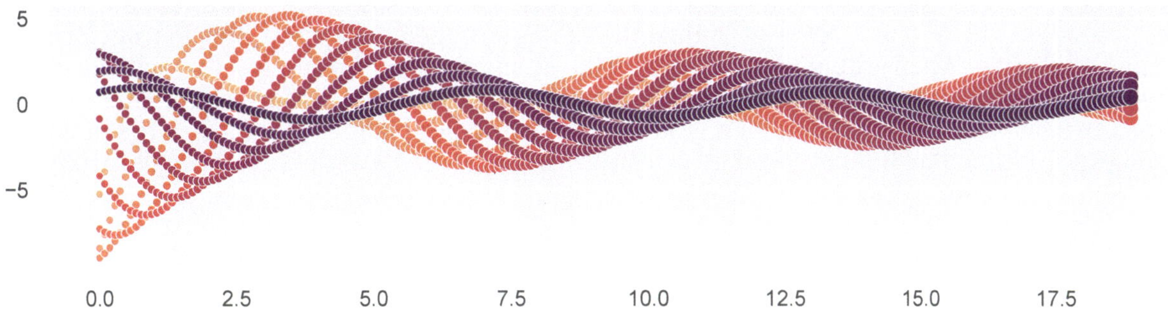

Sinusoidal waves are transformations of sine. While sine and cosine functions have a regular period of 2π and a range from -1 to 1, they can be transformed into sinusoidal waves with different amplitudes, periods, and phases. The basic form of a sinusoidal wave is:

$$y = a\sin(bx - c) + d \qquad \textbf{Equation } 9.2$$

where:

a is amplitude,

b transforms the period $\frac{2\pi}{|b|}$,

c transforms the phase shift $\frac{c}{b}$, and

d is vertical shift.

Sinusoidal waves can be transformed by operations such as amplification, modulation, shifting, scaling, cutting, dampening, and texturing. In the basic form of the sinusoidal wave, a amplifies, b modulates, c translates horizontally, and d translates vertically. A decay factor will dampen a wave, while the sum of sinusoidal functions will texture a wave. Adapting Joseph Choma's approach to mathematical transformations for designers [7], this section explores how different operations transform sinusoidal waves.

Grasshopper To graph a sinusoidal wave in Grasshopper, start by composing the expression `a*sin(b*x-c)+d`. Create a `Range` of values from 0 to 2π and plug them into the expression as its `x` variable. Add number sliders for the variables `a`, `b`, `c`, and `d`. Add a `Construct Point` component, plug the range in as the x-coordinates, and plug the results of the expression in as the y-coordinates. Finally `Interpolate` a curve through the points.

Code: `waves.gh`

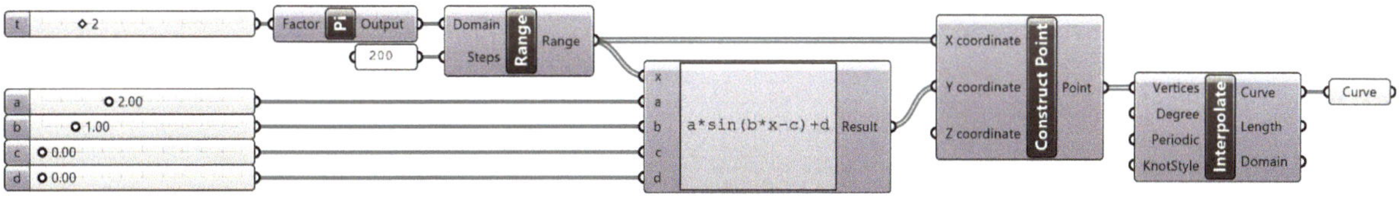

Python To graph a sinusoidal wave in Python, first import the necessary libraries, then set the variables, generate a sequence of x-values, compute y as a function of x, and finally plot the x- and y-values as coordinates. Assign default values to the variables `t`, `steps`, `a`, `b`, `c`, and `d`. Use the function `y = a * np.sin(b * x - c) + d`. Adjust the values assigned to each variable to transform the sinusoidal wave. When graphing the wave with the function `seaborn.scatterplot`, style the graph by setting the size and hue parameters to `x`, the color palette to `'flare'`, and the legend to `False`. See the Seaborn documentation for a guide to choosing color palettes [2]. To save the plot to an image file, assign the plot produced by `seaborn.scatterplot` to a variable, return an instance of the figure containing the plot, set the size of the figure, and then save the figure. When setting the file name, choose a supported type of image such as `.png`, `.pdf`, or `.svg` and a resolution in dots per inch.

Code: `sinusoidal-wave.py`

```
# import libraries
import numpy as np
import seaborn as sns
sns.set_theme()

# set variables
steps = 200
t = 6
a = 2
b = 2
c = 0
d = 0

# plot sinusoidal wave
x = np.linspace(0, t * np.pi, steps)
y = a * np.sin(b * x - c) + d
plot = sns.scatterplot(x=x, y=y, size=x, hue=x, palette='flare', legend=False)
fig = plot.get_figure()
fig.set_size_inches(8.5, 2)
fig.savefig('sine-wave.png', dpi=300)
```

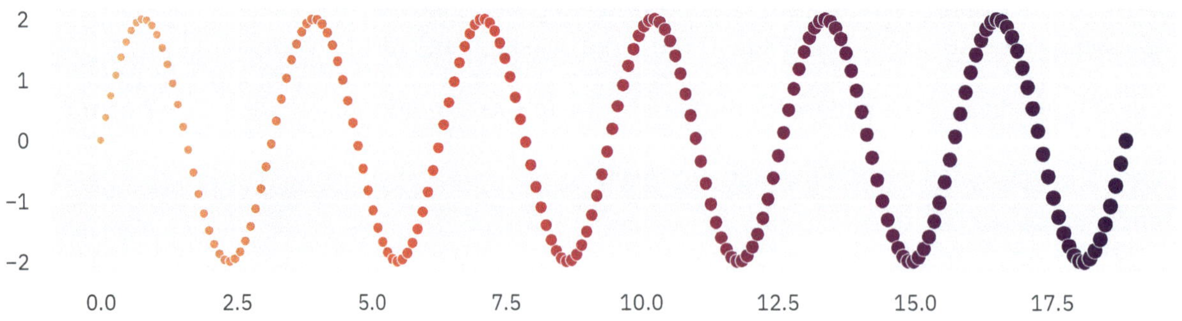

9.3.1 Amplification

Grasshopper To amplify a sinusoidal wave in Grasshopper, calculate the sine of sequence x and then multiply by amplitude a. To implement this in Grasshopper, start by composing a definition for a sine wave. Then add a `Multiplication` component, connect its A input to the output of the `Sine` function and connect its B input to a number slider for amplitude a. Plug `Multiplication`'s output into `Construct Point`'s y-coordinate input.

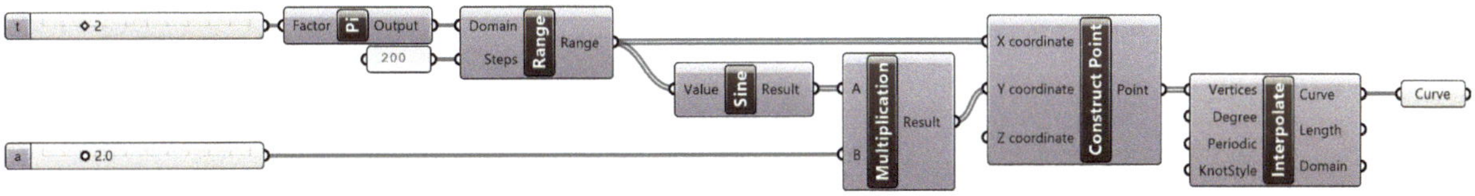

Python Use the equation `y = a * np.sin(x)` to amplify a sinusoidal wave in Python.

Code: `amplified-wave.py`

```
# import libraries
import numpy as np
import seaborn as sns
sns.set_theme()

# set variables
steps = 100
t = 2
a = 2

# plot sinusoidal wave
x = np.linspace(0, t * np.pi, steps)
y = a * np.sin(x)
sns.scatterplot(x=x, y=y, size=x, hue=x, palette='flare', legend=False)
```

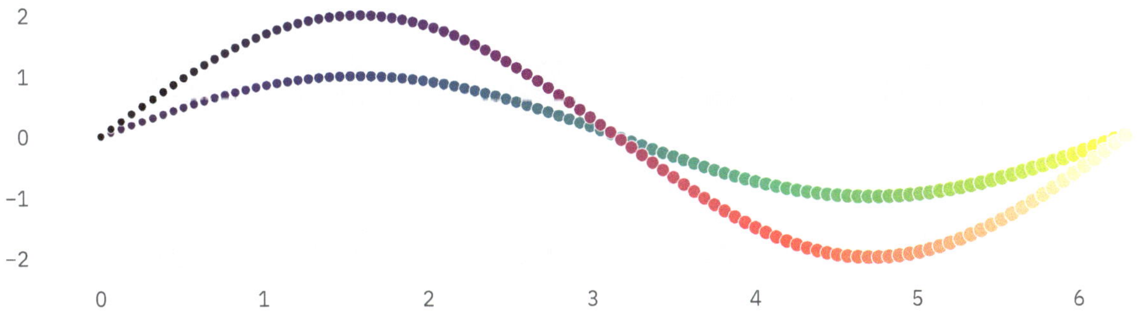

9.3.2 Modulation

Grasshopper To modulate a sinusoidal wave, multiply sequence `x` by `b` before calculating its sine. To implement this in Grasshopper, start by composing a definition for a sine curve. Then add a `Division` component, connect the output of the `Range` to division's input A, and connect a number slider `b` to division's input B. Plug division's output into the `Sine` function's input.

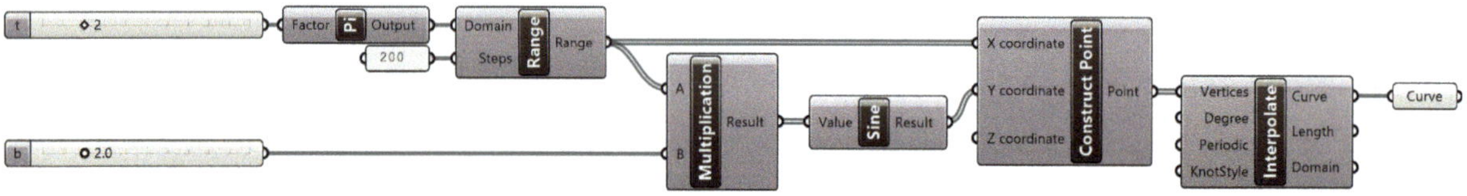

Python Use the equation `y = np.sin(b * x)` to modulate a sinusoidal wave in Python.

Code: `modulated-wave.py`

```
# import libraries
import numpy as np
import seaborn as sns
sns.set_theme()

# set variables
steps = 100
t = 2
b = 2

# plot sinusoidal wave
x = np.linspace(0, t * np.pi, steps)
y = np.sin(b * x)
sns.scatterplot(x=x, y=y, size=x, hue=x, palette='flare', legend=False)
```

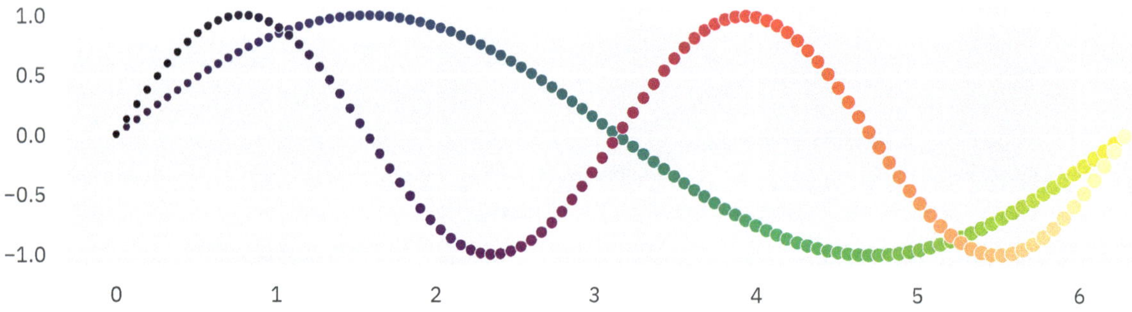

9.3.3 Shifting

Grasshopper To shift a sine curve along the x-axis, subtract c from sequence x, calculate the sine of the sequence, and then add d. To implement this in Grasshopper, start by composing a definition for a sine curve. Then add a `Subtraction` component, connect its A input to the output of the `Range` component, connect its B input to a number slider c, and plug its output result into the `Sine` component. Then add an `Addition` component, plug its A input into the output of the sine function, plug its B input into a number slider d, and plug its output into the y-coordinates of `Construct Point`.

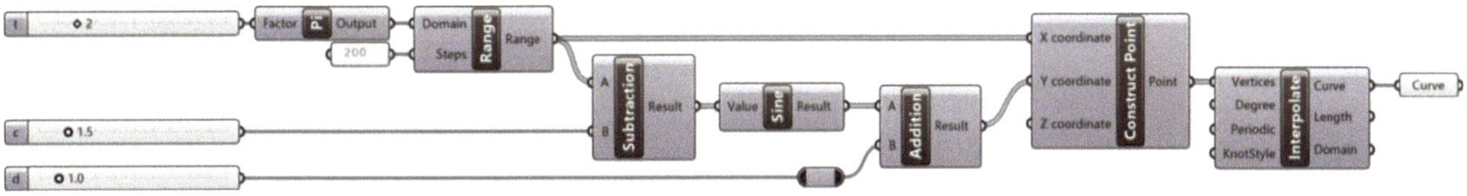

Python Use the equation `y = np.sin(x - c) + d` to translate a sinusoidal wave along the x-axis in Python.

Code: `translated-wave.py`

```
# import libraries
import numpy as np
import seaborn as sns
sns.set_theme()

# set variables
steps = 100
t = 2
c = 1.5
d = 1

# plot sinusoidal wave
x = np.linspace(0, t * np.pi, steps)
y = np.sin(x - c) + d
sns.scatterplot(x=x, y=y, size=x, hue=x, palette='flare', legend=False)
```

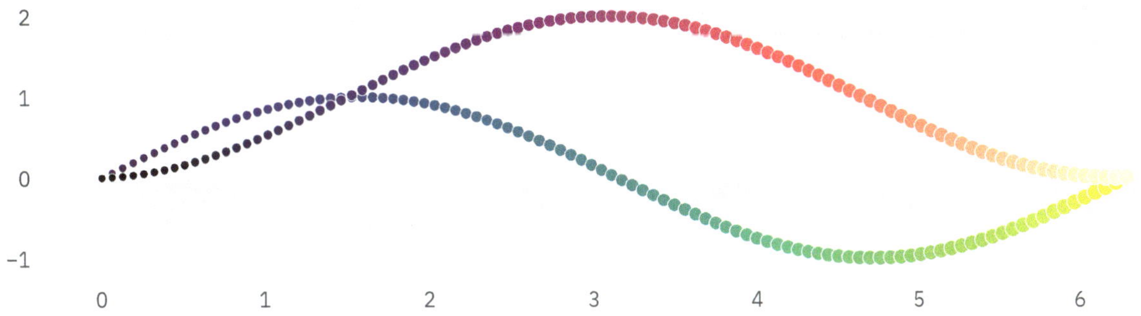

9.3.4 Scaling

Grasshopper To scale a sine curve uniformly, multiply the values for x- and y-coordinates by a scaling factor. To scale it non-uniformly, multiply the coordinates by different scaling factors. Compose a definition for a sine curve. Add a `Multiplication` component, connect its `A` input to the output of the `Range` component, connect its `B` input to a number slider, and plug its output result into the x-coordinate of the `Construct Point` component. Add another `Multiplication` component, connect its `A` input to the output of the `Sine` component, connect its `B` input to a number slider, and plug its output result into the y-coordinate of the `Construct Point` component. For uniform scaling use the same number slider `e` for both multiplications.

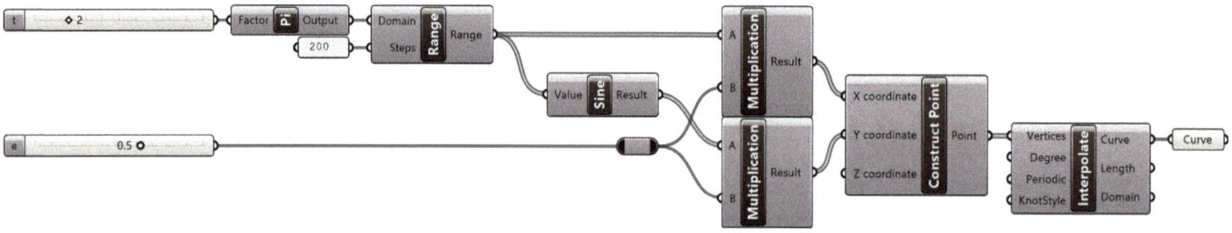

Python To scale a sinusoidal wave in Python, multiply the x- and y-coordinates by a scale factor `e`. When plotting set `x = x * e` and `y = y * e`.

Code: `scaled-wave.py`

```
# import libraries
import numpy as np
import seaborn as sns

# set variables
steps = 100
t = 2
e = 0.5

# plot sinusoidal wave
x = np.linspace(0, t * np.pi, steps)
y = np.sin(x)
sns.scatterplot(x=x * e, y=y * e, size=x, hue=x, palette='flare', legend=False)
```

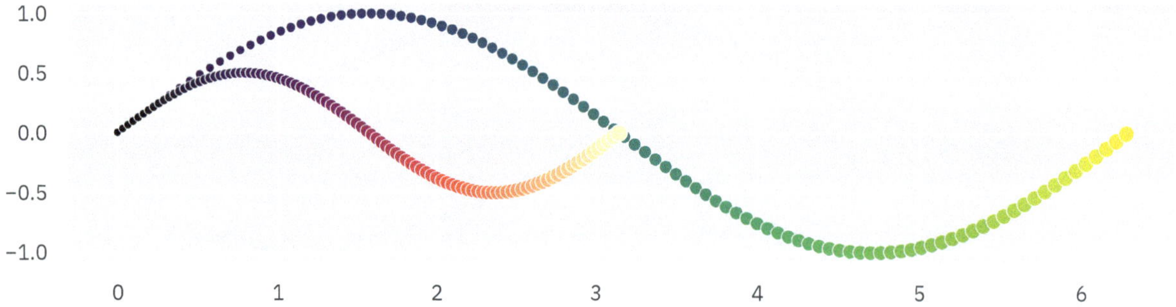

9.3.5 Cutting

Grasshopper To cut or extend a sine curve, simply change the end of its domain by factor of π. In Grasshopper, compose a definition for a sine curve and then adjust the `t` number slider connected to the `Pi` and `Range` components.

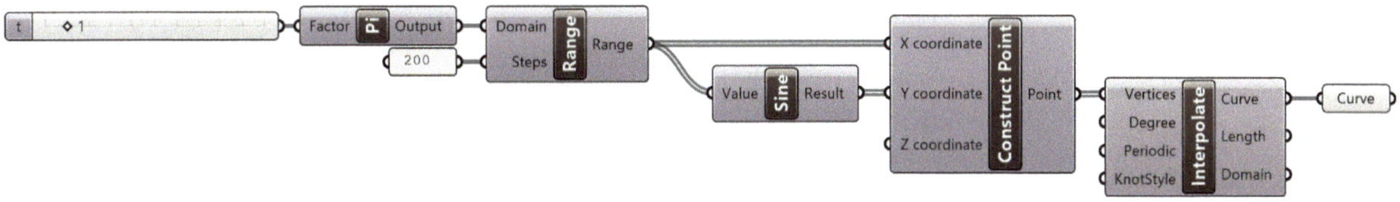

Python To cut a sinusoidal wave in Python, just adjust variable `t`. Consider scaling the `steps` proportionally.

Code: `cut-wave.py`

```
# import libraries
import numpy as np
import seaborn as sns
sns.set_theme()

# set variables
steps = 50
t = 1

# plot sinusoidal wave
x = np.linspace(0, t * np.pi, steps)
y = np.sin(x)
sns.scatterplot(x=x, y=y, size=x, hue=x, palette='flare', legend=False)
```

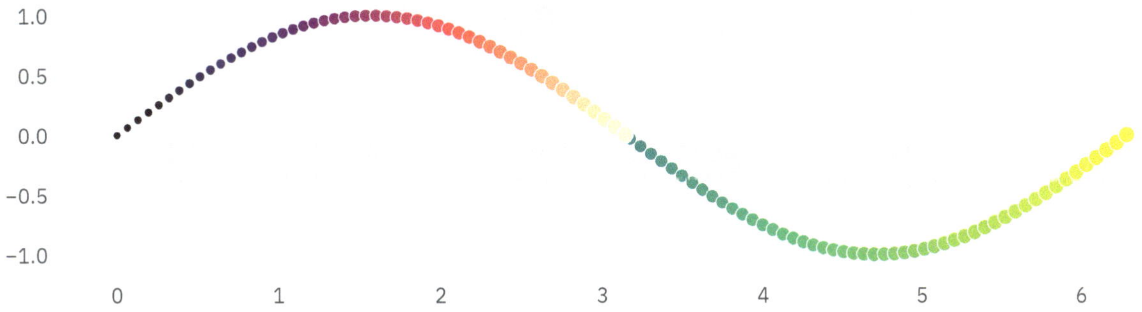

9.3.6 Dampening

To dampen a sinusoidal wave multiply the sine of sequence x by a decay function.

dampening

$$\{x \mid 0 \leq x \leq 4\pi\}$$

$$y = a \cdot \sin(b \cdot x) \cdot e^{-\lambda \cdot x}$$

dampening series

$$\{x \mid 0 \leq x \leq 4\pi\}$$

$$y = \sum_{i=m}^{n} a \cdot \sin(b \cdot x) \cdot e^{-\lambda_i \cdot x}$$

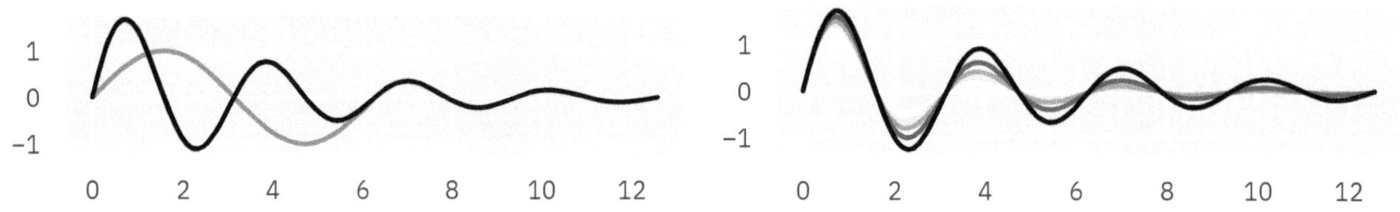

Figure 9.3: Dampened waves

Grasshopper To implement a sinusoidal wave dampened by the exponential decay function $f(x) = e^{-\lambda x}$ in Grasshopper, start by composing a definition for a sine wave. Then start to compose the decay function by adding a `Power of E` component. Connect its input exponent to the output of a `Multiplication` component. For the `Multiplication` component, connect its A input to the output of the `Range` component and connect its B input to a `Negative` component connected to a number slider `f` for the decay constant λ. Finally add a `Multiplication` component, connect its A input to the output of the `Sine` component, connect its B input to the output of the decay function, and plug its output result into the y-coordinate of the `Construct Point` component.

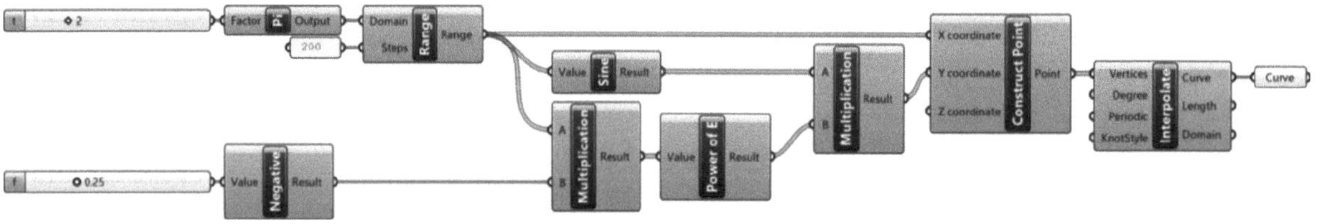

Expressions are useful when composing more complicated equations. Use expressions when performing compound transformations on trigonometric waves. To modulate, amplify, and dampen a sine wave, start by composing an `Expression` for a sine curve. Then edit the expression so that it reads `a*sin(b*x)*e^(-f*x)`. Add the variables x, a, b, and `f` as input parameters for the expression. To add additional input parameters to a component, zoom in until a plus button appears, click the plus button, and then right click to rename the parameter.

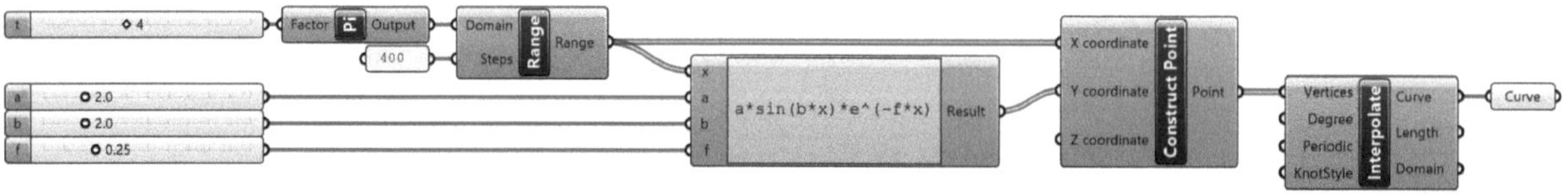

To generate dampened sinusoidal waves that decay at incrementally greater rates in Grasshopper, compose an expression for dampening a wave and then graft in a series of decay constants `f`. Each decay constant will be its own branch of a data tree. Rather than generating a single list of y-coordinates, the expression will generate a list of y-coordinates for each branch of the data tree.

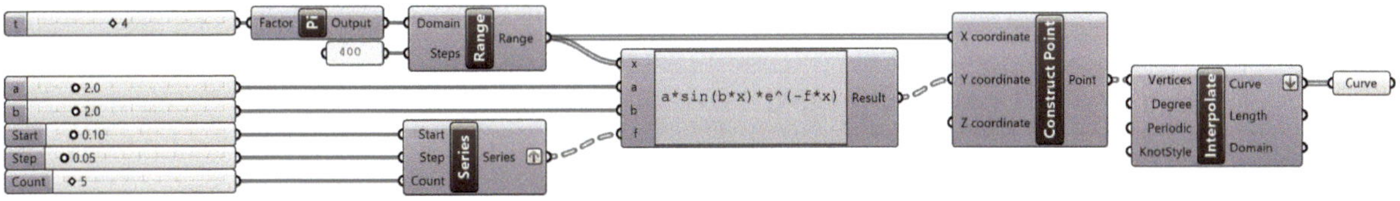

Python

Python To dampen a sinusoidal wave in Python, use the equation `y = a * np.sin(b * x) * np.e**(-f * x)`. The decay function `np.e**(-f * x)` will cause the amplitude of the wave to decrease over time.

Code: `dampened-wave.py`

```
# import libraries
import numpy as np
import seaborn as sns
sns.set_theme()

# set variables
steps = 100
t = 4
a = 2
b = 2
f = 0.25

# plot sinusoidal wave
x = np.linspace(0, t * np.pi, steps)
y = a * np.sin(b * x) * np.e**(-f * x)
sns.scatterplot(x=x, y=y, size=x, hue=x, palette='flare', legend=False)
```

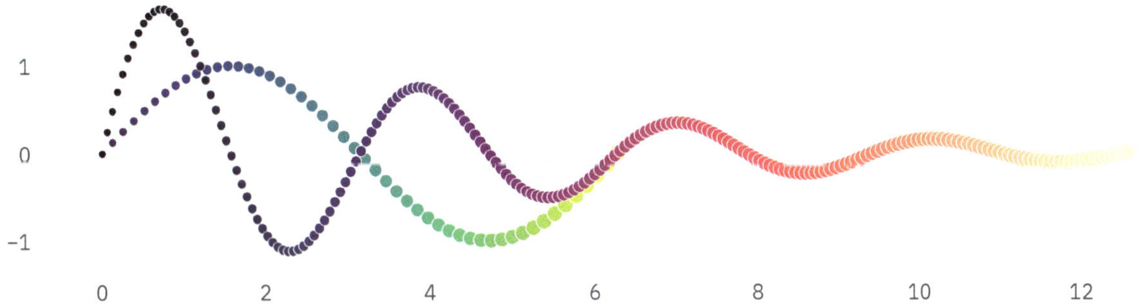

Graph a sequence of incrementally dampened sinusoidal waves. After generating a sequence of x-values, generate a sequence of decay constants λ with `numpy.linspace`. Iterate through the array of decay constants with a **for** loop. Solve the function `y = a * np.sin(b * x) * np.e**(-i * x)` for each decay constant `i`. Set the hue of the scatterplot to `i` and normalize the hue from the minimum to maximum decay constants in the sequence.

Code: `dampened-waves.py`

```
# import libraries
import numpy as np
import seaborn as sns
sns.set_theme()

# set variables
steps = 200
count = 10
minimum = 0.2
maximum = 0.6
t = 4
a = 2
b = 2

# plot sinusoidal waves
x = np.linspace(0, t * np.pi, steps)
f = np.linspace(maximum, minimum, count)
for i in f:
    y = a * np.sin(b * x) * np.e**(-i * x)
    sns.scatterplot(
        x=x,
        y=y,
        size=x,
        hue=i,
        hue_norm=(minimum, maximum),
        palette='magma',
        legend=False
        )
```

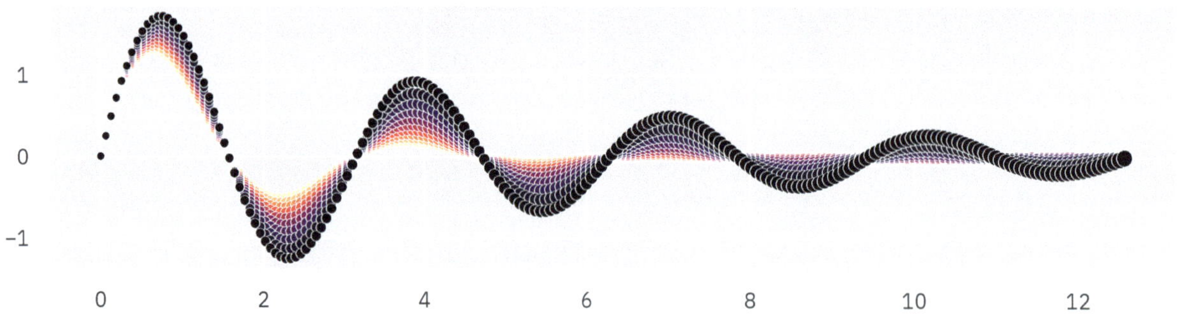

9.3.7 Texturing

Adding sinusoidal functions creates a textured wave. When waves coincide, they become superimposed – their forces are added together. The resulting wave is the sum of their displacement.

texturing

$$\{x \mid 0 \leq x \leq 2\pi\}$$

$$y = a \cdot \sin(b \cdot x) + \sin(x)$$

texturing series

$$\{x \mid 0 \leq x \leq 2\pi\}$$

$$y = \sum_{i=m}^{n} a_i \cdot \sin(b \cdot x) + \sin(x)$$

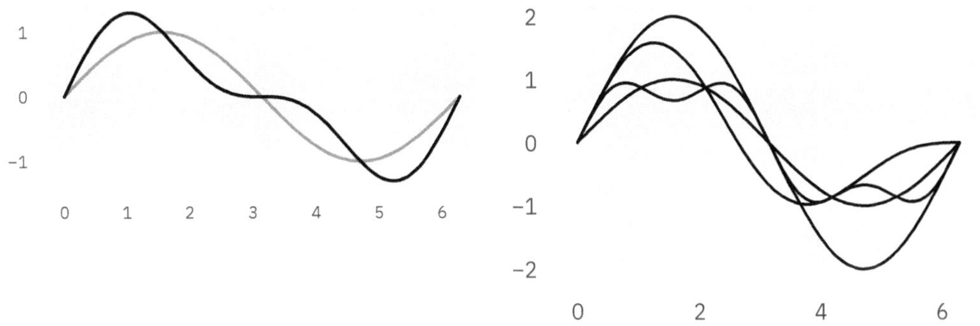

Figure 9.4: Textured waves

Grasshopper To texture a wave, add sinusoidal functions of differing frequencies. To implement this in Grasshopper, compose an expression for a sine wave in Grasshopper. Then edit the expression so that it reads `a * sin(a^-1 * x) + sin(x)` with b as the inverse of a. Add the variables x and a.

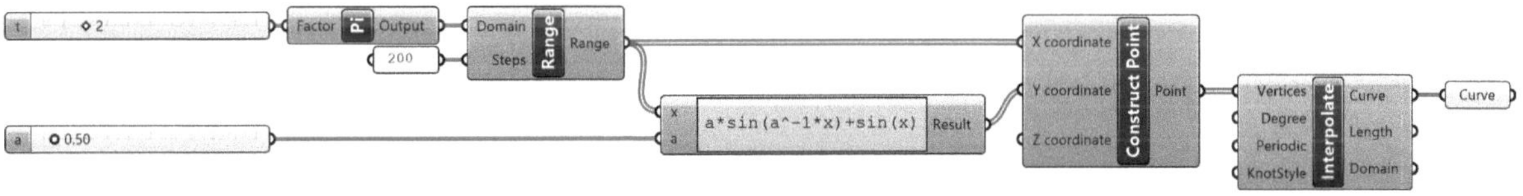

To generate a series of textured waves in Grasshopper, compose the expression `a * sin(a^-1 * x) + sin(x)` and graft in a series of amplitudes a.

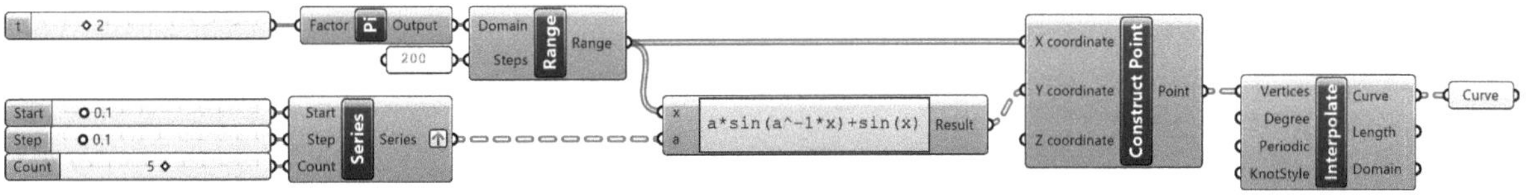

Python Use the equation `y = a * np.sin(a**-1 * x) + np.sin(x)` to texture a sinusoidal wave in Python.

Code: `textured-wave.py`

```
# import libraries
import numpy as np
import seaborn as sns
sns.set_theme()

# set variables
steps = 100
t = 2
a = 0.5

# plot sinusoidal wave
x = np.linspace(0, t * np.pi, steps)
y = a * np.sin(a**-1 * x) + np.sin(x)
sns.scatterplot(x=x, y=y, size=x, hue=x, palette='flare', legend=False)
```

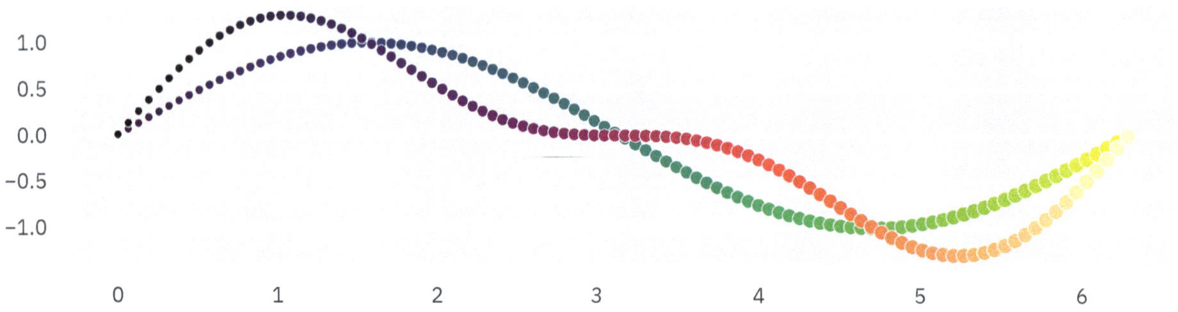

Graph a sequence of textured sinusoidal waves. After generating a sequence of x-values, generate a sequence of amplitude values with `numpy.linspace`. Iterate through the array of amplitudes with a **for** loop. Solve the function `y = i * np.sin(i**-1 * x) + np.sin(x)` for each amplitude `i` with `b` as the inverse of `i`. To style the sequence of waves, set the hue of the scatterplot to `i` and normalize the hue from the minimum to maximum amplitude values in the sequence.

Code: `textured-waves.py`

```
# import libraries
import numpy as np
import seaborn as sns
sns.set_theme()

# set variables
steps = 100
count = 4
minimum = 0.001
maximum = 1.000
t = 2

# plot sinusoidal waves
x = np.linspace(0, t * np.pi, steps)
a = np.linspace(minimum, maximum, count)
for i in a:
```

```
y = i * np.sin(i**-1 * x) + np.sin(x)
sns.scatterplot(
    x=x,
    y=y,
    size=x,
    hue=i,
    hue_norm=(minimum, maximum),
    palette='flare',
    legend=False
    )
```

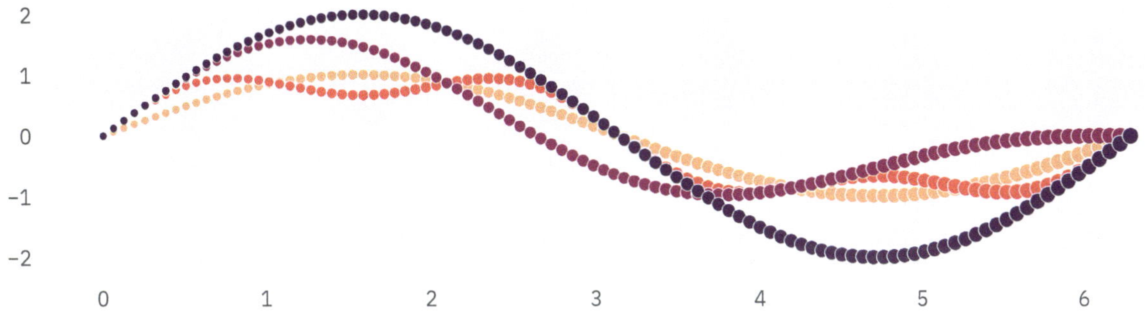

References

[1] [Software] NumPy Developers, *NumPy* version 1.25.0, 2023. lic: BSD 3-Clause License. url: https://numpy.org, vcs: https://github.com/numpy/numpy.

[2] Michael Waskom. *Choosing color palettes*. 2022. url: https://seaborn.pydata.org/tutorial/color_palettes.html (visited on 08/12/2023).

[3] [Software] Michael Waskom, *Seaborn: statistical data visualization* version 0.12.2, 2022. lic: BSD 3-Clause License. url: https://seaborn.pydata.org, vcs: https://github.com/mwaskom/seaborn.

[4] Michael L. Waskom. "seaborn: statistical data visualization". In: *Journal of Open Source Software* 6.60 (2021), page 3021. doi: 10.21105/joss.03021.

[5] Charles R. Harris, K. Jarrod Millman, Stéfan J. van der Walt, Ralf Gommers, Pauli Virtanen, David Cournapeau, Eric Wieser, Julian Taylor, Sebastian Berg, Nathaniel J. Smith, Robert Kern, Matti Picus, Stephan Hoyer, Marten H. van Kerkwijk, Matthew Brett, Allan Haldane, Jaime Fernández del Río, Mark Wiebe, Pearu Peterson, Pierre Gérard-Marchant, Kevin Sheppard, Tyler Reddy, Warren Weckesser, Hameer Abbasi, Christoph Gohlke, and Travis E. Oliphant. "Array programming with NumPy". In: *Nature* 585.7825 (Sept. 2020), pages 357–362. doi: 10.1038/s41586-020-2649-2.

[6] Snøhetta. *MAX IV Laboratory Landscape*. 2016. url: https://snohetta.com/projects/70-max-iv-laboratory-landscape (visited on 07/28/2021).

[7] Joseph Choma. *Morphing: A Guide to Mathematical Transformations for Architects and Designers*. Laurence King Publishing, 2015. isbn: 9781780674131.

[8] Mode Lab. *The Grasshopper Primer*. 3rd edition. Robert McNeel & Associates, 2015. url: http://grasshopperprimer.com.

[9] Maya Lin. *Storm King Wavefield*. 2009. url: https://www.mayalinstudio.com/art/storm-king-wavefield (visited on 06/04/2022).

[10] Orencio Durán, Veit Schwämmle, and Hans J. Herrmann. "Breeding and solitary wave behavior of dunes". In: *Physical Review E* 72 (2 Aug. 2005), page 021308. doi: 10.1103/PhysRevE.72.021308.

[11] Hicham Elbelrhiti, Philippe Claudin, and Bruno Andreotti. "Field evidence for surface-wave-induced instability of sand dunes". In: *Nature* 437 (7059 2005), pages 720–723. issn: 1476-4687. doi: 10.1038/nature04058.

[12] Maya Lin. *The Wave Field*. 1995. url: https://www.mayalinstudio.com/art/the-wave-field (visited on 06/04/2022).

[13] George Gabriel Stokes. "On the Theory of Oscillatory Waves". In: *Transactions of the Cambridge Philosophical Society* 8 (1847), pages 441–455.

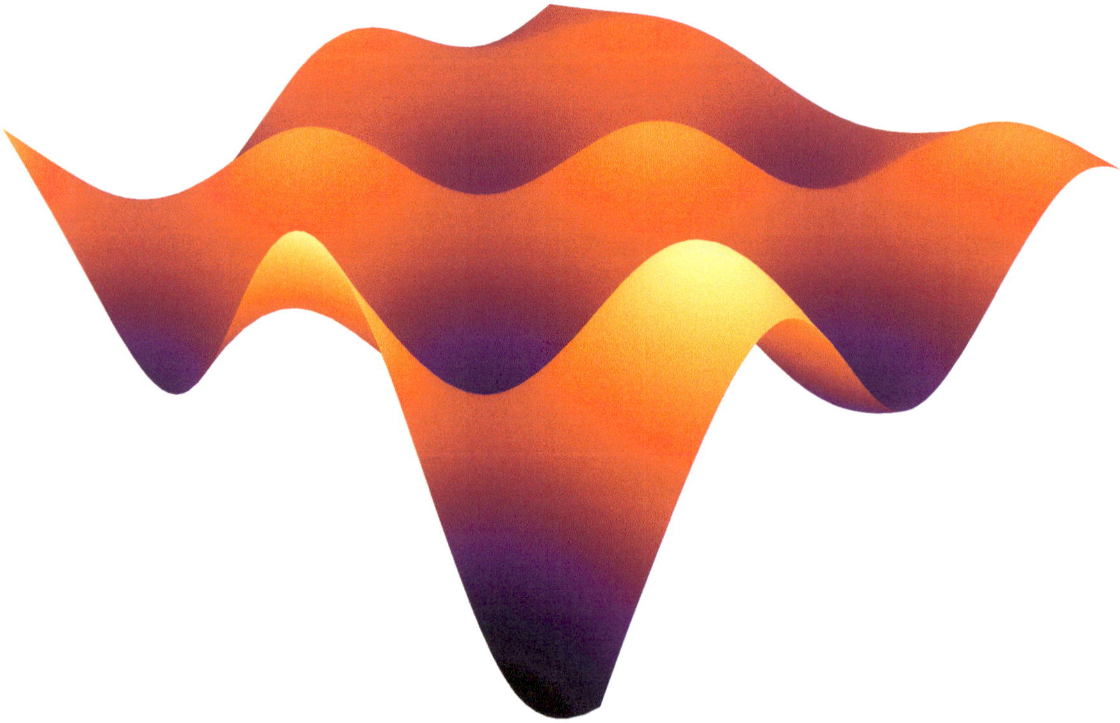

10 Surfaces

This chapter explores how to plot 3-dimensional sinusoidal waves as points, surfaces, contours, and voxels. By generating such surfaces landscape architects can model topographic and ecological gradients.

10.1 Surfaces in Grasshopper

Surface Plot a 3-dimensional sinusoidal wave with the form $z = \sin(x)\cos(y)$ first as a set of points and then as a surface. Start by generating a sequence of numbers with the `Series` component. This sequence of numbers will be used to construct a grid of points. First use this list of numbers to construct the x- and y-coordinates of a grid of points. Plug the output of `Series` into the x- and y-coordinates of `Construct Point`. Graft the x-coordinate input so that the resulting array is 2-dimensional. Next construct the z-coordinates as a function of the x- and y-coordinates. Create the expression `sin(x)*cos(y)` and plug the sequence of numbers into the expression as its x and y variables. Graft the x variable so that the expression operates on the 2-dimensional array of x- and y-coordinates. This will generate a 3-dimensional wave of points. Use the `Surface from Points` component to interpolate a surface through this set of points. Flatten the input tree of points into a list, set the u count to the number of elements in the `Series`, and set interpolate to `True` with either a Boolean toggle or a panel.

Code: `wave-surface.gh`

$$z = \sin(x)\cos(y) \qquad \textbf{Equation } 10.1$$

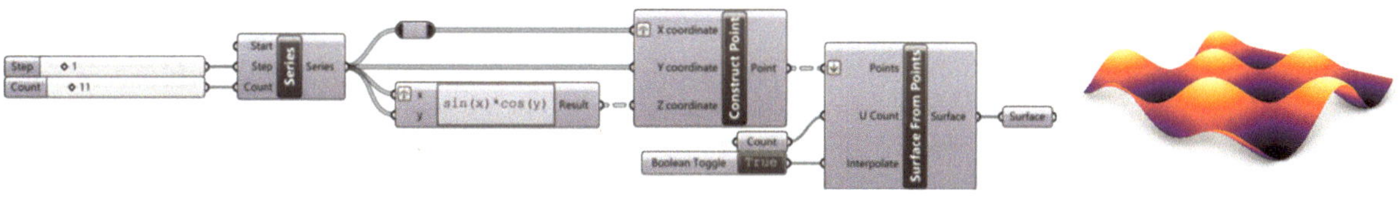

Expression Try constructing a more complicated expression such as:

```
(a * sin(b * x - c) + d) * e^(-f * x) * (a * cos(b * y - c) + d) * e^(-f * y)
```

$$z = (a\sin(b\,x - c) + d)\,e^{-\lambda_i x}\,(a\cos(b\,y - c) + d)\,e^{-\lambda_i y} \qquad \textbf{Equation } 10.2$$

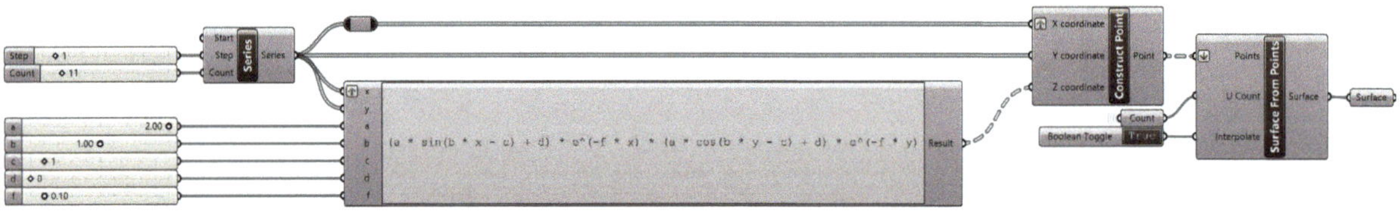

Color Gradient To assign a color gradient to the surface, convert the surface into a mesh and color the vertices of the mesh by their height z. First use the `Mesh Brep` component to convert the surface into a mesh. Then use the `Deconstruct Mesh` component to extract the vertices of the mesh. Use the `Deconstruct Point` component to extract the z-component of each vertex. Use the `Remap Numbers` component to remap the domain of the list of z-values to 0 to 1. Use the list of z-components as `Remap Numbers`' input value and the `Bounds` of the z-components as the source domain. Plug the mapped numbers into a gradient as the input parameter and set a color ramp. Finally, use a `Mesh Colours` component to assign the color gradient to the mesh.

Grasshopper's color gradient has a limited palette of predefined color maps. Double click on the component to create your own customized color map. Alternatively use a custom gradient from `color-maps.gh` or try the Grasshopper add-on `Color+` [1], which has a large collection of color gradients including perceptually uniform color maps such as cividis, viridis, inferno, magma, plasma, and cubehelix. Perceptually uniform color maps do not distort the data they represent and are accessible for those with color-vision deficiency or color-blindness [3]. See appendix B for a guide to color gradients.

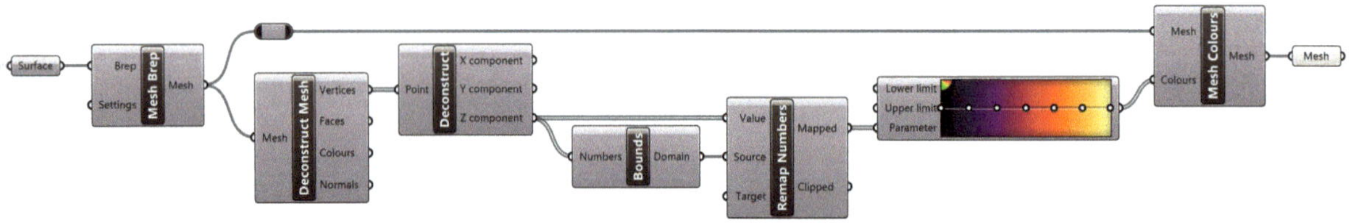

Contours Compute contours for the surface. Start by adding the `Contour` component. Set its input shape to the surface and set its input distance with a number slider. Flatten the output data tree of contours into a list. To color the contours by elevation, add a `Custom Preview`, set its geometry to the list of contours, and set the material to a `Gradient`. Set `Gradient`'s input parameter to a `Range`. Set `Range`'s input steps to the number of contours using `List Length`.

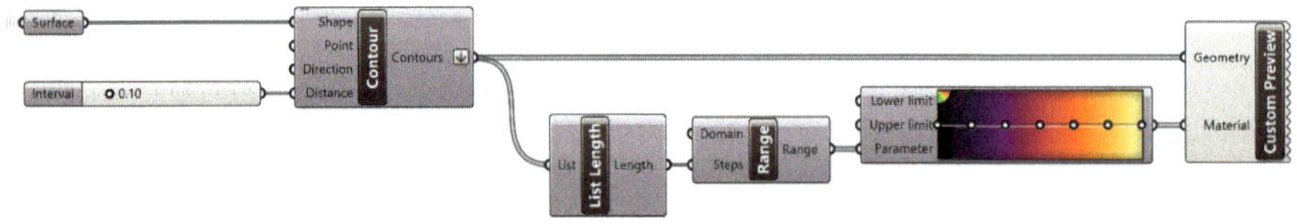

10.2 Surfaces in Python

Matplotlib is a Python library for data visualization [2, 4]. Use Matplotlib's `mplot3d` toolkit to generate 3D plots such as scatterplots, surfaces, contours, and voxels with Python.

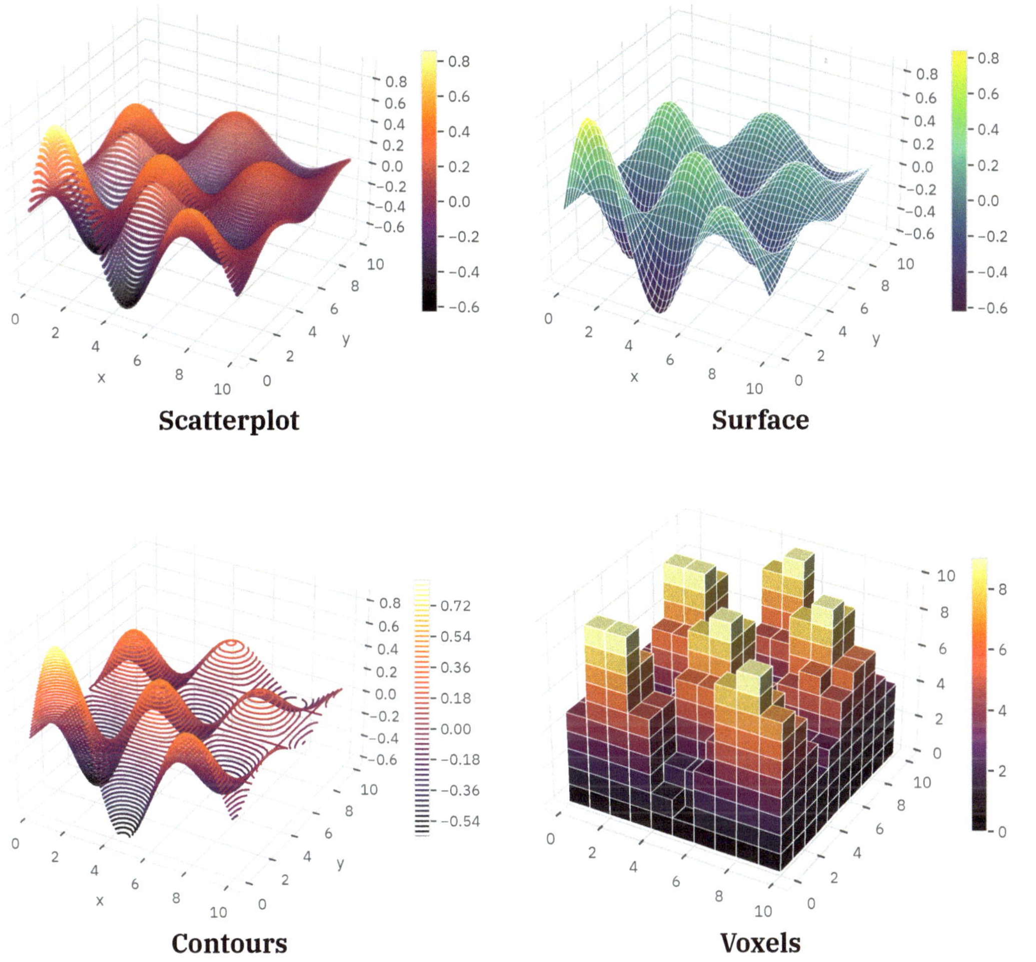

Figure 10.1: 3D plots

10.2.1 3D Scatterplots

A 3-dimensional scatterplot is a simple way to plot a surface such as a wave. A 3-dimensional scatterplot is a type of point cloud – a set of points defined by x-, y-, and z-coordinates. Use NumPy to generate a set of x-, y-, and z-coordinates from a sinusoidal wave and Matplotlib to plot these as a scatterplot.

Code: `wave-scatter.py`

First import the necessary libraries, then instantiate the plot with a 3-dimensional projection, generate a set of x-, y-, and z-coordinates from a sinusoidal wave, plot these coordinates as a scatterplot, set labels on the plot axes, and create a legend. Start by importing NumPy and Matplotlib's `pyplot` interface. Optionally import Seaborn and set the plot style to `whitegrid` with the function `seaborn.set_style`. Create a plot with the function `matplotlib.pyplot.subplot`, set the projection to 3d, and assign it to an object ax. Create one-dimensional arrays with numpy.arange and assign them to the variables u and v. Then set the variables a, b, c, d, and f for the sinusoidal function. Create a grid with numpy.meshgrid whose coordinates are defined by the arrays u and v. Assign the coordinate matrices of the grid to the variables x and y. Then compute z with the function: z = np.sin(x) * np.cos(y). Plot the coordinates as a 3D scatterplot using the function Axes3D.scatter. Set the x-, y-, and z-coordinates, set the color sequence to the z-coordinates with c=z, set the point size with s=5, and set the color map with cmap='inferno'. For a perceptually uniform color map that represents topography try, 'gist_earth', 'terrain', or 'cubehelix'. Then set labels for each axis of the plot, retrieve the figure containing the plot, and create a legend with matplotlib.pyplot.colorbar. After successfully plotting, try a more complicated function such as: z = ((a*np.sin(b*x-c)+d)*np.e**(-f*x)*(a*np.cos(b*y-c)+ d)*np.e**(-f*y)).

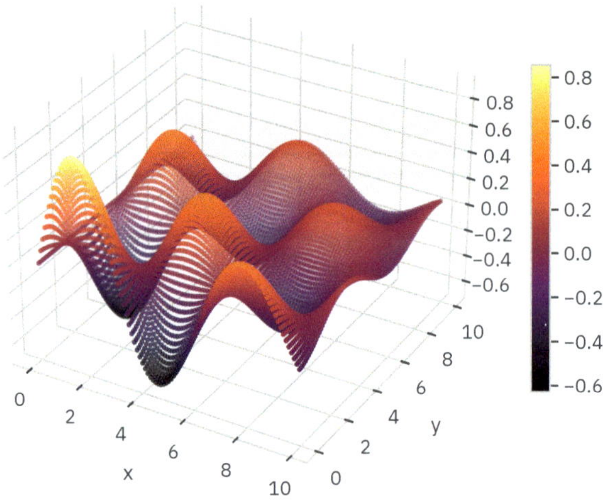


```
# import libraries
import numpy as np
import matplotlib.pyplot as plt
import seaborn as sns
sns.set(style = "whitegrid")

# create plot
ax = plt.subplot(projection='3d')

# set variables
u = np.arange(0, 10, 0.1)
v = np.arange(0, 10, 0.1)
a = 1
b = 1
c = 0
d = 0
f = 0.1

# plot points
x, y = np.meshgrid(u, v)
z = (
    (a * np.sin(b * x - c) + d)
    * np.e**(-f * x)
    * (a * np.cos(b * y - c) + d)
    * np.e**(-f * y)
    )
plot = ax.scatter(
    x, y, z,
    c=z,
    s=5,
    cmap='inferno'
    )

# set axes label
ax.set_xlabel('x', labelpad=10)
ax.set_ylabel('y', labelpad=10)
ax.set_zlabel('z', labelpad=10)

# create legend
fig = ax.get_figure()
fig.colorbar(plot, shrink=0.5, aspect=16)
```

10.2.2 Surfaces

To plot a surface instead of a scatterplot, replace the function `Axes3D.scatter` with `Axes3D.plot_surface`. Set the line width of the surface patches with the `linewidths` parameter.

Code: `wave-surface.py`

```
plot = ax.plot_surface(
    x, y, z,
    cmap='viridis',
    linewidth=0.25
    )
```

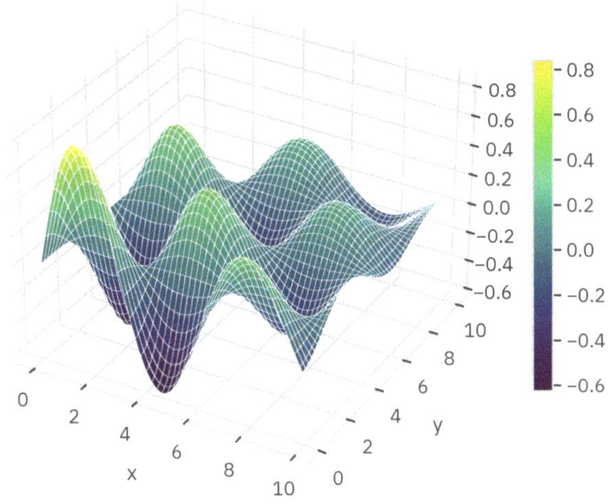

10.2.3 Contours

To plot contours instead of a scatterplot, replace the function `Axes3D.scatter` with `Axes3D.contour`. Set the number of contours with the `levels` parameter and the width of the contours with the `linewidths` parameter.

Code: `wave-contours.py`

```
plot = ax.contour(
    x, y, z,
    levels=50,
    linewidths=1.5,
    cmap='inferno'
    )
```

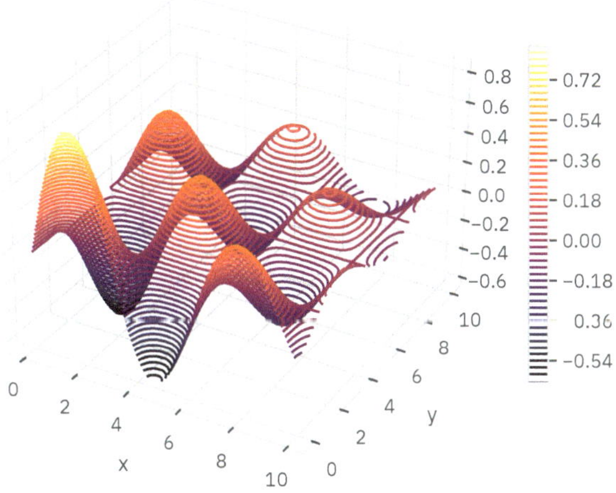

10.2.4 Voxels

Voxels – volumetric pixels – are 3-dimensional grids, i.e. rasters. This data structure is especially useful for representing, modeling, and processing point clouds. A surface such as a wave function can easily be plotted as voxels with Matplotlib's `Axes3D.voxels` function. Rather than directly plotting a wave function as coordinates, create a 3-dimensional matrix of coordinates and then find the z-coordinates underneath the wave function. First generate a grid of x-, y-, and z-values with `numpy.indices`. Use an inequality to find z-coordinates under a surface – here, a 3D wave. Assign the resulting 3D array of Boolean values (i.e. true or false) to a variable named `volume`. Create a plot with the function `matplotlib.pyplot.subplot`, set the projection to `3d`, and assign it to an object `ax`. To create a color gradient from the z-values, first normalize these values to a domain from 0.0

to 1.0 with the function `matplotlib.colors.Normalize`. Then assign the normalized z-values to a color map. Use the normalized color map for the color of the faces when plotting the voxels. Plot the data with `Axes3D.voxels`. Set the first parameter – the cells to fill – to the 3D array of Boolean values assigned to the variable `volume`. Then set the color of the faces to the normalized color map, set the color of the edges, and set the line width. Finally create a legend with a normalized color bar using `matplotlib.pyplot.colorbar`.

Code: `wave-voxels.py`

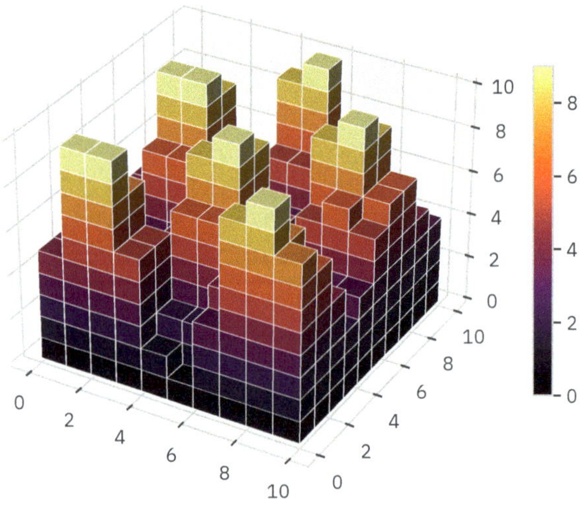


```
# import libraries
import numpy as np
import matplotlib as mpl
import matplotlib.pyplot as plt
import seaborn as sns
sns.set(style = "whitegrid")

# set variables
u = 10
v = 10
w = 10

# plot voxels
x, y, z = np.indices((u, v, w))
volume = (
    z < w/2
    * np.sin(x)
    * np.cos(y)
    + w/2
    )
ax = plt.subplot(projection='3d')
norm = mpl.colors.Normalize(
    vmin=z.min(),
    vmax=z.max()
    )
colors = plt.cm.inferno(norm(z))
ax.voxels(
    volume,
    facecolors=colors,
    edgecolor='w',
    linewidth=0.5
    )

# create legend
legend = plt.cm.ScalarMappable(
    cmap=plt.cm.inferno,
    norm=norm
    )
legend.set_array([])
plt.colorbar(
    legend,
    shrink=0.5,
    aspect=16,
    ax=plt.gca()
    )
```

References

[1] [Software] David Mans, *Color+* version 1.5.0, 2023. lic: MIT License. url: https://interopxyz.gitbook.io/color-plus/, vcs: https://github.com/interopxyz/ColorPlus.

[2] [Software] Matplotlib Development Team, *Matplotlib* version 3.7.2, 2023. lic: Matplotlib License. url: https://matplotlib.org, vcs: https://github.com/matplotlib/matplotlib.

[3] Fabio Crameri, Grace E. Shephard, and Philip J. Heron. "The misuse of colour in science communication". In: *Nature Communications* 11.1 (2020), page 5444. issn: 2041-1723. doi: 10.1038/s41467-020-19160-7.

[4] John D. Hunter. "Matplotlib: A 2D Graphics Environment". In: *Computing in Science and Engineering.* 9.3 (June 2007), pages 90–95. issn: 1521-9615. doi: 10.1109/MCSE.2007.55.

11 Patterns

In this example, a set of sinusoidal waves generate an interwoven network of pathways and planting. New permutations of this system of paths can be generated simply by changing the parameters of the wave function. Use the Grasshopper definition `planting-pattern.gh` as a guide for this example. For a variant with polygonal trees, try `lowpoly-planting.gh`.

Data: `quercus-alba-01-s.3dm`

Code: `planting-pattern.gh`

11.1 Input

In Rhino, open a tree from the point cloud library such as `quercus-alba-01-s.3dm`, `.e57`, or `.xyz`. In Grasshopper, add a point cloud parameter. Then right click on this parameter and set it to the tree in the Rhino scene.

11.2 Paths

Waves Create a path system from sinusoidal waves. Start by plotting a sequence of sinusoidal waves with the form $y = a \cdot \sin(b \cdot x) + \sin(x)$ as a set of curves. In Grasshopper compose the `expression` `a * sin(a^-1 * x) + sin(x)`. Set input `x` to a sequence from 0 to 2π and graft in a series of amplitudes for input `a`. Then construct a set of points with the `Construct Point` component. Set the input x-coordinate to the sequence from 0 to 2π and the input y-coordinate to the output of the expression. Interpolate a curve through this set of points with an `Interpolate` component.

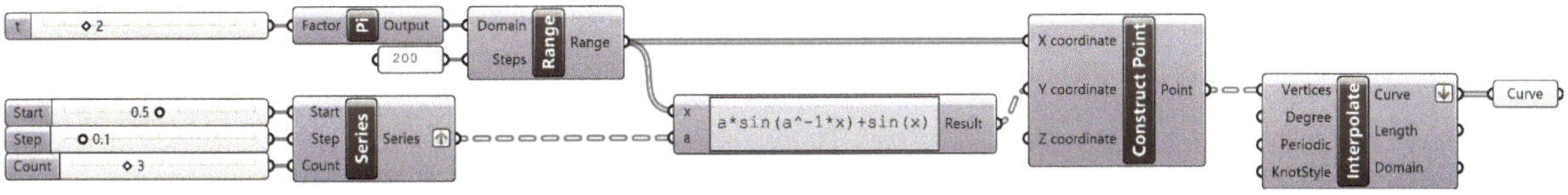

Paths Model a network of paths from these curves. Start by scaling this set of curves to the size of path desired with a `Scale` component. This initial set of curves will act as the left edges of the paths. Offset these curves with an `Offset Curve` component to create the right edges of the paths. Flatten the resulting data tree into a list of curves. Use a `Ruled Surface` component to generate surfaces between the left and right edges of the paths. Merge the resulting surfaces with a `Region Union` component. Then create a planar surface from the resulting set of closed curves with a `Boundary Surfaces` component.

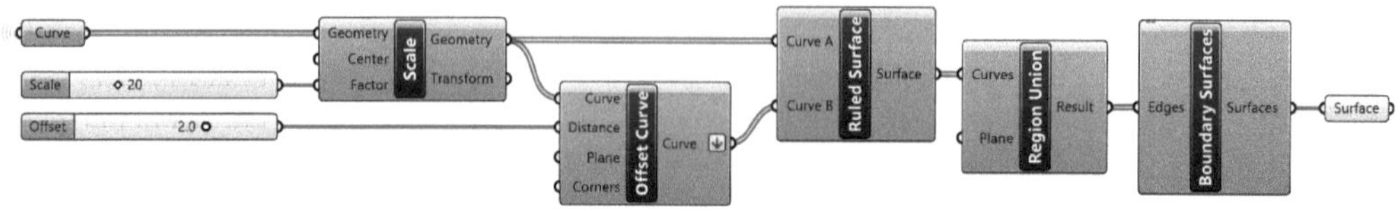

Extrusion Extrude the surface of the path into a solid. Use an `Extrude` component to give the path a thickness. Set the direction of extrusion with a `Unit Z` component and a number slider. Use a `Custom Preview` with a `Colour Swatch` assigned as its material to render the extrusion.

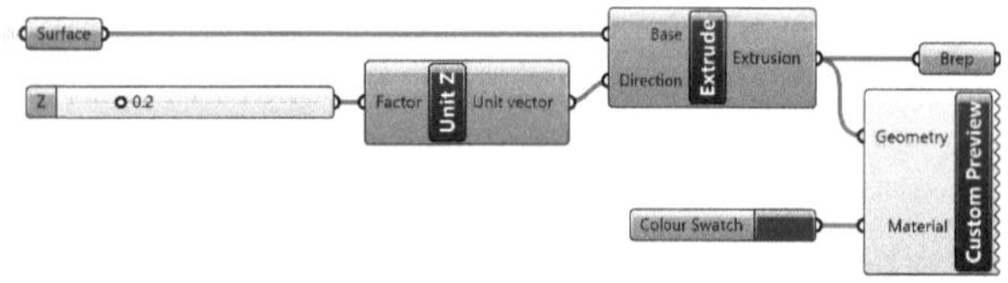

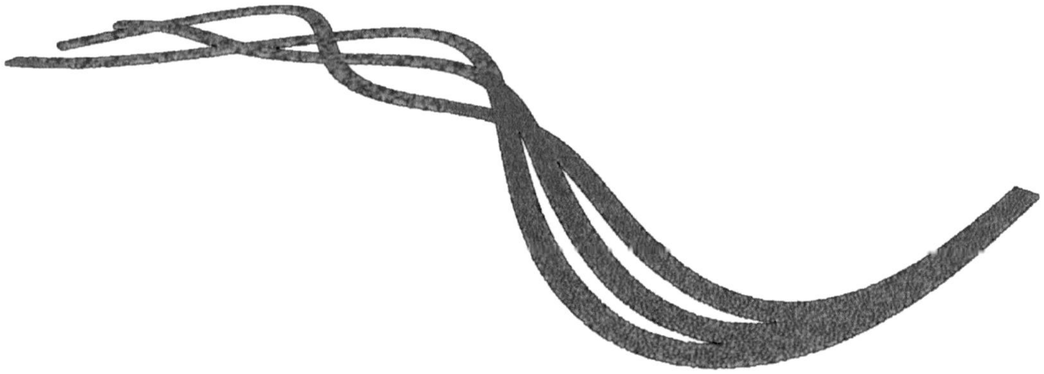

11.3 Trees

Waves Distribute trees along another set of sinusoidal waves. Use the expression from section 11.2 with different input parameters to plot a new set of curves.

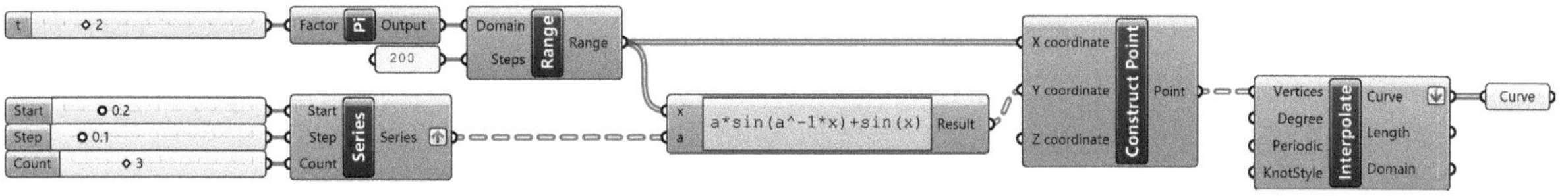

Positioning Site the trees at regular intervals along these curves. First, scale the curves to real-world dimensions with a `Scale` component. Then divide the curves by a set distance with a `Divide Distance` component. This will generate sequences of regularly spaced points along each curve. Flatten this data tree into a list and remove points that are too close with a `Cull Duplicates` component.

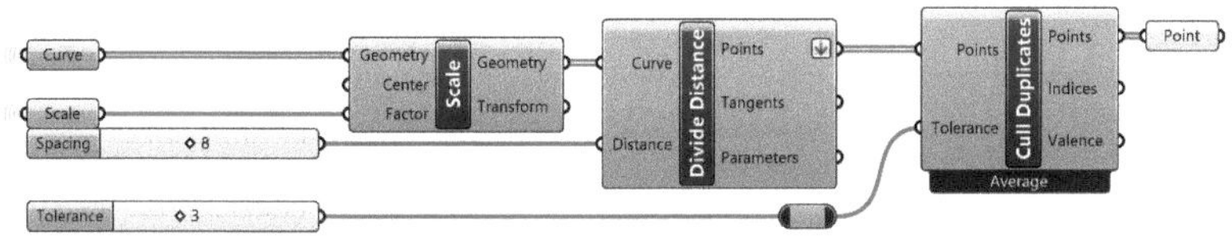

Culling Remove planting locations that cross the paths. First determine which planting locations intersect the paths using a `Point in Brep` component. Set the input boundary representation (BRep) to the extruded path and set the input points to the planting locations. This component will report whether points are inside the solid or not with a list of true or false values. Create a culling pattern by inverting the true and false values using Boolean negation with a `Gate Not` component. Then use a `Cull Pattern` component to remove points based on this pattern. Find the number of remaining points with a `List Length` component.

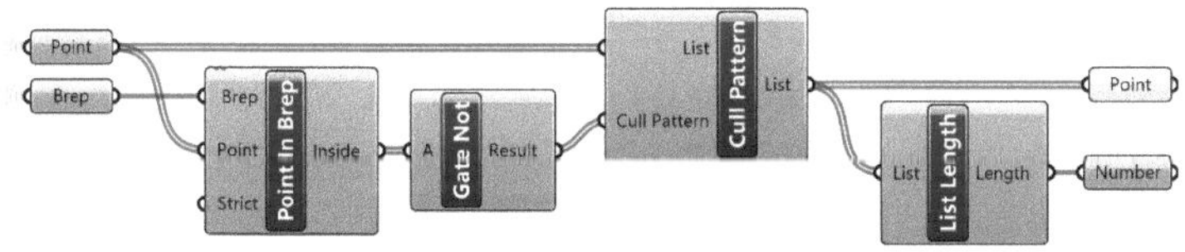

Translation Move a copy of the point cloud model to each remaining point. Add a `Move` component, set the input geometry to the point cloud, and set the motion using a `Vector 2Pt` component. Create a vector between the origin at `0,0,0` and the list of points.

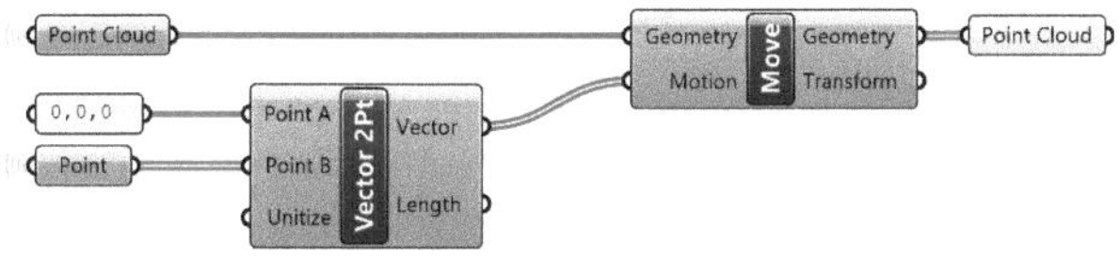

Scaling Randomly scale the trees. First generate a list of random scale factors with a `Random` component. Set the input range to `0.7 To 1` and set the number of random values to the number of remaining points calculated earlier with `List Length`. Then scale the point clouds using a `Scale` component. Set the center of scaling to the list of remaining points and the factor to the random list.

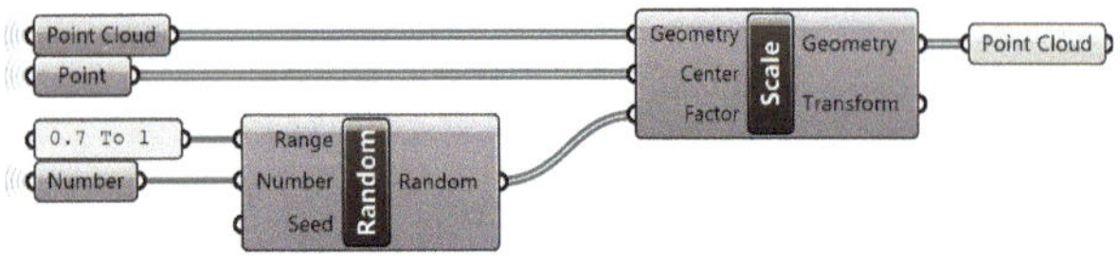

Rotation Randomly rotate the trees. Generate a random list of rotation angles with a `Random` component with a range from `0 To 360`. Then rotate the point clouds using a `Rotate` component. Right click on the input angle parameter and set it to degrees, rather than radians. Set the angle to the random list and set the rotation plane to the list of remaining points.

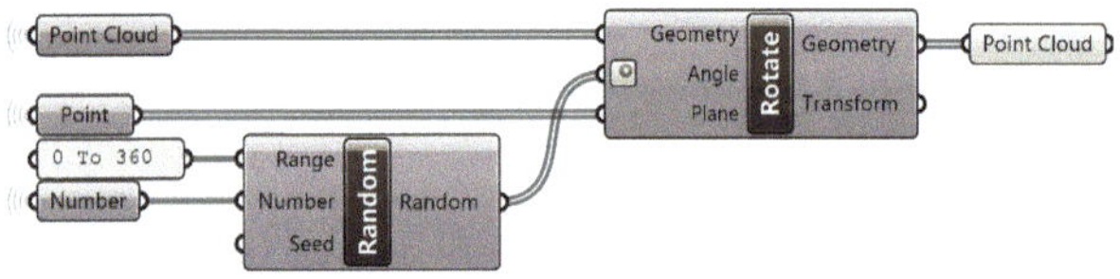

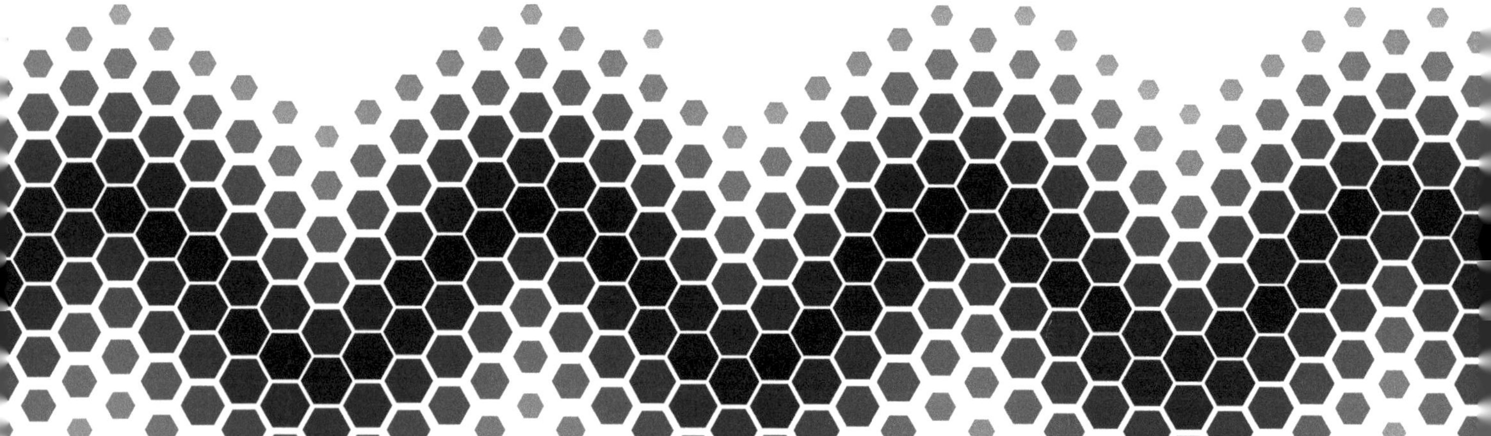

12 Attractors

In this example a sinusoidal wave acts as an attractor for paving – creating a sinuous pathway with a gradient of pavers. The further the pavers are from the wave, the smaller and lighter they become. Attractors are fields of attraction or repulsion. They can be used to transform geometry, cull objects, or map colors along gradients. Attractors can be used by landscape architects to design gradients of plants, paving, or landforms.

Code: `wave-attractor.gh`

12.1 Sinusoidal Wave

Start by plotting a sequence of points from a sinusoidal wave with the form $y = a\ \sin(bx - c) + d$.

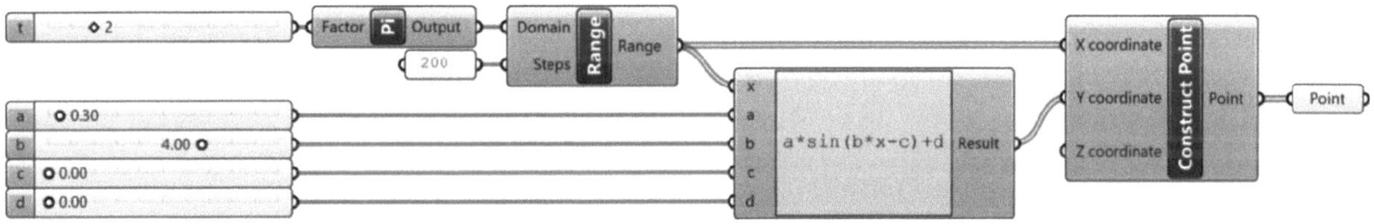

12.2 Attractor

The points plotted by the wave will act as an attractor for paving, scaling, culling, and coloring a grid of pavers. First generate a grid with the `Hexagonal` component. Flatten the output trees of cells and points into lists. Compute how far each point of the grid is away from the wave with the `Closest Point` component. Set the flattened list of grid points as the input point parameter and set the list of points from the wave as the input cloud parameter. Map this distance field to a new domain using the `Remap Numbers` component. Set `Remap Numbers`' input value parameter to the list of distances generated by `Closest Point`. Use the `Bounds` component to find the domain of the list of distances and plug that in as the input source parameter. Set the input target parameter to a new domain such as `0.85 To 0.1` defined by a panel. Note that the domain should go from larger to smaller, so that the pavers decrease in size the further they are from the wave. Try using the graph mapper to rescale the output list of mapped numbers non-linearly. Place the graph mapper on the canvas, input the list of the mapped numbers, right click on the graph mapper to set the type of graph to a `Bezier` curve, and adjust the handles of the curve to rescale the values. Use this list of numbers to `Scale` the grid cells generated by the `Hexagonal` component. Place the `Scale` component, set its input geometry to the flattened list of cells, set its center of scaling to the flattened list of points, and set its input factor to the output list of numbers from the graph mapper. Each hexagonal cell will be scaled by the distance of its centroid to the wave.

Use a `Cull Pattern` component to remove pavers that are far from the wave. First test whether the list of numbers generated by the graph mapper is `Larger Than` a threshold. This will generate a list of Boolean values, either true or false. Then place a `Cull Pattern` component, plug `Scale`'s output geometry in as its input list, and plug the list of Boolean values from the `Larger Than` component in as its input cull pattern. Try changing the parameters of the wave and the attractor to generate different paving patterns.

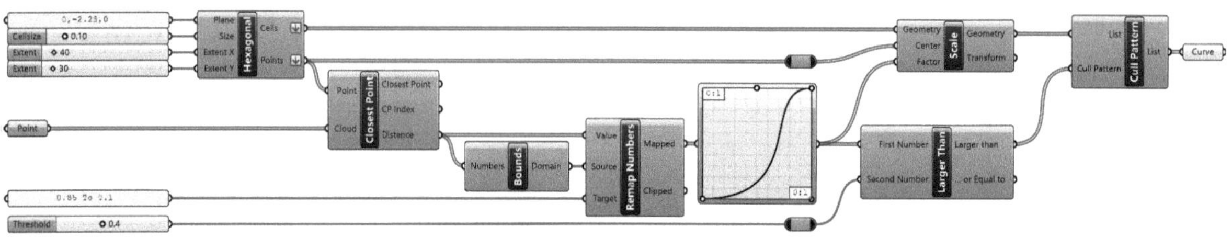

12.3 Styling

To style the pavers, model each cell as a surface and map a color gradient across the surfaces. The `Boundary Surfaces` component generates surfaces from closed planar curves. Use this component to model the scaled, culled list of cells as surfaces. Add a `Custom Preview` component, plug the list of boundary surfaces in as its input geometry, and plug a gradient in as its input material. Right click on the gradient to set the color table to a preset such as grayscale or double click on it to customize the color table. To represent the distance of the pavers from the wave, the gradient should be parameterized with the list of remapped and culled distance values. To do this, plug the list of numbers output by the graph mapper in as the input list for a `Cull Pattern`. Set its cull pattern to `Larger Than`'s output list of Booleans. Then set gradient's lower limit to the threshold used by the `Larger Than` component and set its parameter input to the culled list.

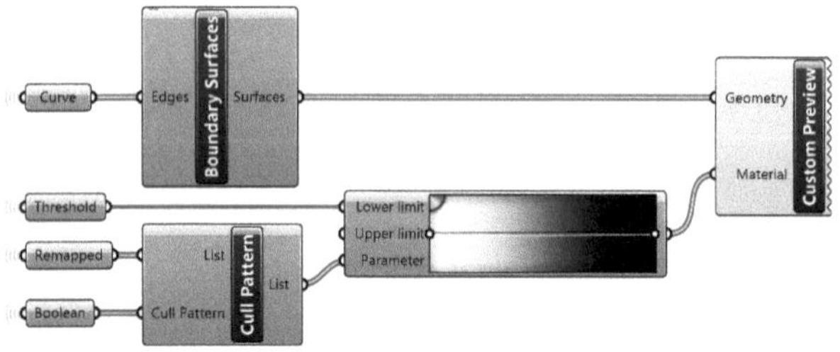

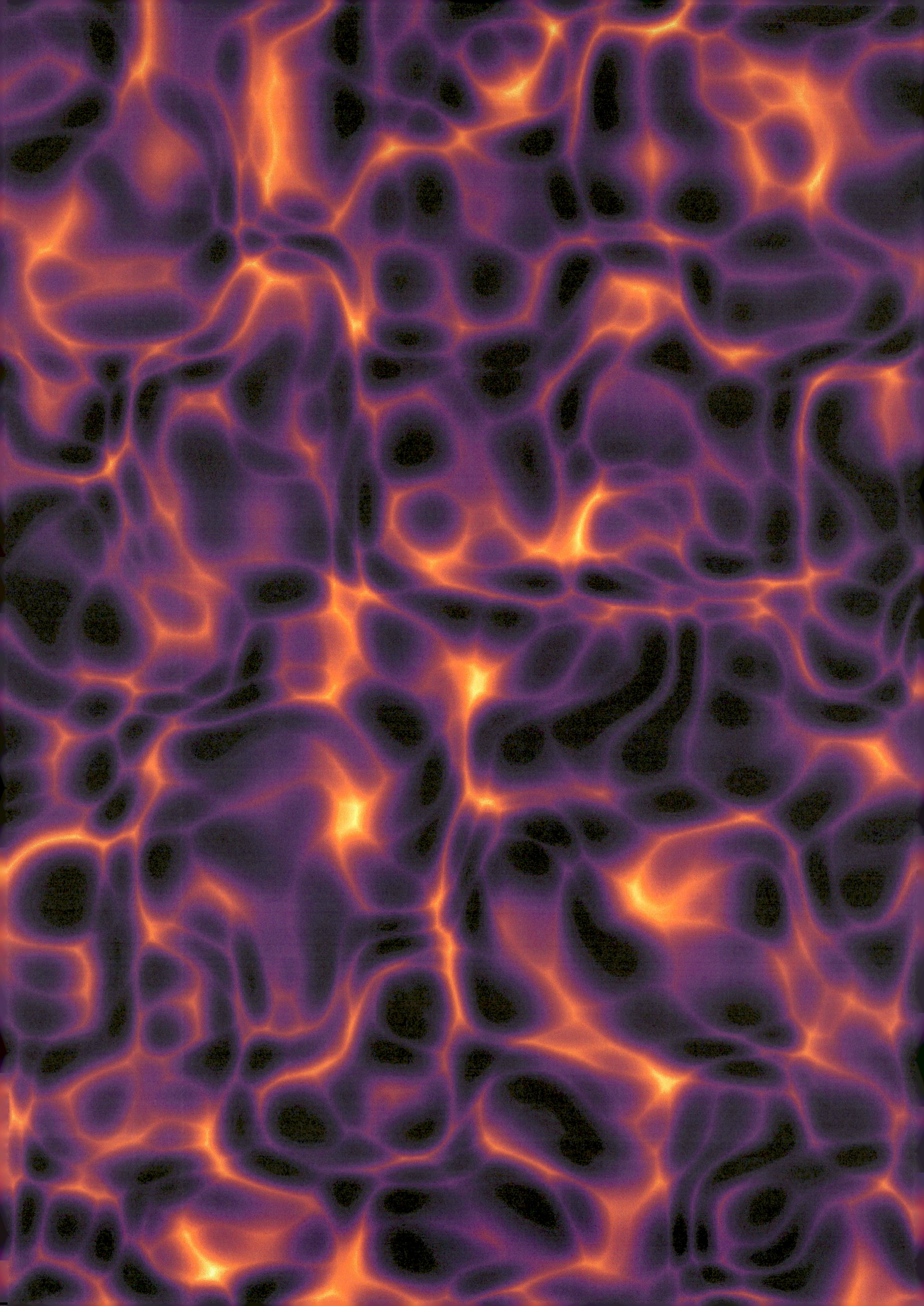

13 Noise

13.1 Procedural Noise

Noise is the imperceptible; acoustic waves beyond audible frequencies, for example, are background noise. Noise is confused signals; audible, but indecipherable acoustic waves are noise rather than sound. Noise is unwanted signals such as static on the radio, the crackle of a vinyl record, the jitter of poor digital recordings, the murmur of voices during a performance or film, the rumble of traffic, or artifacts like the Moiré effect in digital images. Noise, however, is not always undesirable. Background noise gives music context and depth – it is the reason that compact discs sound sterile, while vinyl sounds richer and live performance richer still [12]. Noise has been used in art, cinema, music, architecture, and computer graphics. For example in Semiconductor's experimental film Brilliant Noise [11], artifacts from solar wind and noise storms that are typically edited out of solar observations become music, blurring the boundary between signal and noise [7].

Procedural noise is used in computer graphics to simulate the richness and complexity of nature. At a fine scale, procedural noise is used to generate textures for materials such as wood, stone, concrete, and bare soil. At a landscape scale, fractal noise is used to model terrain, biomes, and atmosphere [19, 13]. Noise is also used to create unnatural, synthetic patterns, forms, and animations that have a novel aesthetic – an aesthetic of the algorithm. In architecture, Perlin noise has been used, for example, by Gramazio Kohler Research to model the undulating brick façade of the Kitrvs winery [5, 4]. With procedural noise, designers can make creative use of randomness, balancing between the accidental and intentional. This chapter uses the FastNoise library [3], which includes algorithms for value, Perlin [20], simplex, cellular [18], and fractal noise with domain warping.

White Noise White noise is simply an array of pseudo-random values. It lacks structure, has no pattern. Value noise is generated by interpolating between an array of pseudo-random values. It appears like blurred randomness, but with a blocky, gridded structure. Because the values are interpolated between grid coordinates, the structure of the grid dominates value noise.

Gradient Noise Gradient noise – such as Perlin and simplex noise – is generated by interpolating between the gradients, i.e. slopes, of a grid of pseudo-random values. Gradient noise appears like smoothly blurred randomness. Smooth, undulating, organic forms arise from stochasticity. In Perlin noise the gradient is interpolated between coordinates in a square or cubic grid using a Hermite $3t^2 - 2t^3$ [20] or quintic $6t^5 - 15t^4 + 10t^3$ [16] blending function. In simplex noise the gradient is interpolated between coordinates in a simplex – triangular or tetrahedral – grid to remove the orthogonal directional bias and resulting artifacts found in Perlin noise [16]. A physical effect similar to gradient noise can be achieved in architecture using fabric-formed concrete or plaster. In Matsys' P_Wall, for example, plaster was poured into a mold of nylon fabric held in tension by a network of algorithmically distributed wooden dowels [10, 8]. Gravity pulled the plaster into smoothly undulating forms with an organic, smoothness resembling gradient noise.

Cellular Noise Cellular noise – also known as Worley or Voronoi noise – divides space into cellular regions [18]. For every point on a grid search for the nth closest point in a set of pseudo-random points. Solve for the distance between the closest, second closest, third closest points and so on. The first-order distance – the distance to the closest point – defines a Voronoi diagram, a set of cellular regions that partition space based on the minimum distance to sampling points. Different textures can be generated, for example, from the first-order distance, the second-order distance, or the sum, difference, product, or quotient of the first- and second-order distances.

Fractal Noise Fractals can add depth and detail to procedural noise. Fractal noise is generated by adding or subtracting iterations of procedural noise together. Brownian motion describes random walks – random movements through space that have self-similar, fractal properties. In fractional Brownian motion (fBm) each step is not completely independent, but rather contingent on past steps. Fractional Brownian motion can be applied to procedural noise by adding together multiple iterations, i.e. octaves, of noise waves with incremental steps in frequency and amplitude [19, 15, 6]. Key properties of fractal noise include octaves, lacunarity, and gain. Lacunarity is a measure of texture, of the size distribution of gaps, of empty space. In fractional Brownian motion, lacunarity determines the incremental change in frequency with each octave. Gain, a function of the Hurst index, determines the incremental change in amplitude with each octave. Billow noise is generated by adding the absolute value of each octave of noise. Ridged multifractal noise is generated by subtracting the inverse of the absolute value of each octave of noise [15]. Domain warping displaces input values, distorting the noise surface. It can be used to model atmospheric turbulence and geomorphic processes [14, 17].

13.2 Noise in Grasshopper

Generate 3D surfaces from procedural noise with Grasshopper using the definition fast-noise.gh as a guide. While there are several Grasshopper plugins for generating procedural noise including 4D Noise [9], this chapter uses Bitmap+ [1], which is powered by the FastNoise Lite library [3].

Plugin: `Bitmap+`

Code: `fast-noise.gh`

13.2.1 Perlin Noise

Noise To create a 3D mesh from Perlin noise, begin by placing a `Procedural Noise` component from Bitmap+. Use number sliders to set a random seed, the width and height of the grid, the depth, and frequency for the procedural noise. Set the mode to `Perlin` and set the interpolation method to either a `Hermite` curve where $f(x) = 3x^2 - 2x^3$ or to a smoother `Quintic` curve where $f(x) = 6x^5 - 15x^4 + 10x^3$. The output image will be treated as a height map – i.e. a digital elevation model – where the value of each pixel is its elevation z. For a quick preview of the image on the canvas add the `Prev Image` component from Bitmap+.

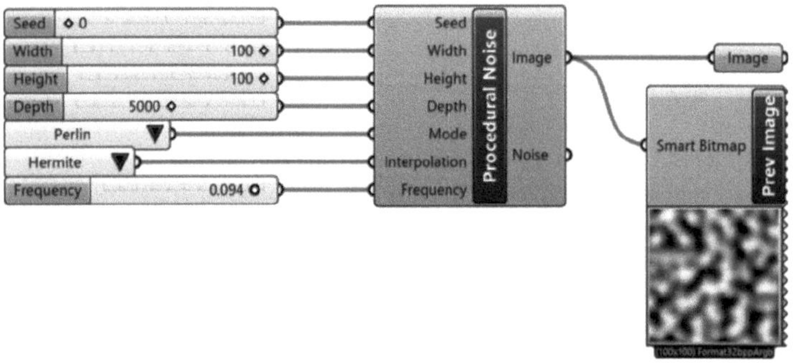

Grid Next generate a mesh from the image grid. To create a mesh, plug the image into a `Bitmap Properties` component to find its boundary, width, and height. Plug these properties into a `Mesh plane` component to generate a mesh grid. Right click on both the input width and height parameters and set the expression `x-1` for each. Then use the `Deconstruct Mesh` component to extract the vertices and faces of the mesh.

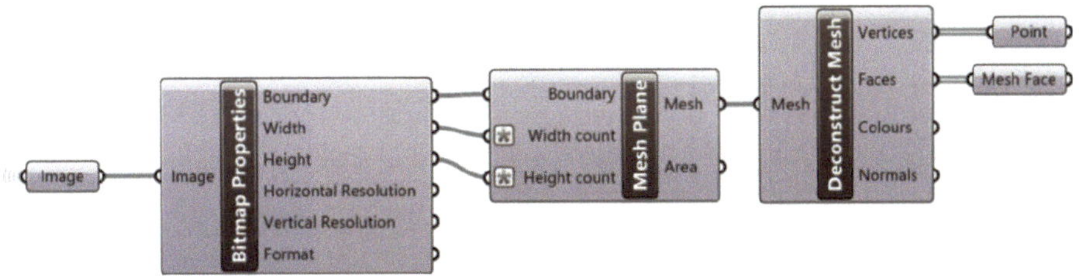

Elevation To extract elevation values from the image, first deconstruct the image into its colors with the `Deconstruct Bitmap` component. Find the value, i.e. the brightness, of colors with the `Split AHSV` component.

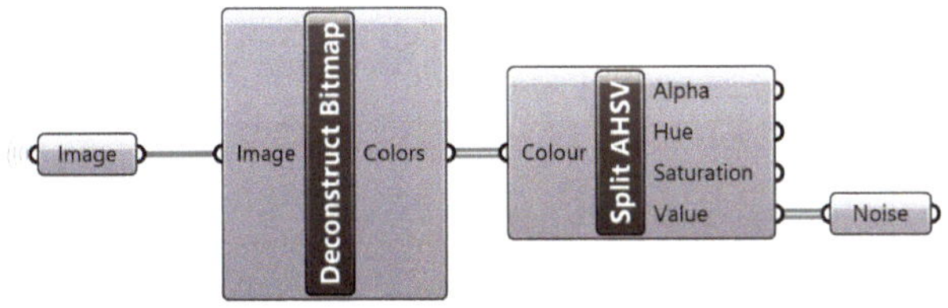

Mesh Use these elevation values to move the vertices of the mesh along the z-axis and a generate a color gradient. Plug the vertices into a `Move` component as the input geometry and set the motion to the `Unit Z` vector. Set the unit vector's factor to the list of values multiplied by a scaling factor. Also plug the list of values into a `Gradient` to generate a color map. Finally, use the `Construct Mesh` component to regenerate the mesh from the translated vertices, the original faces, and the color map.

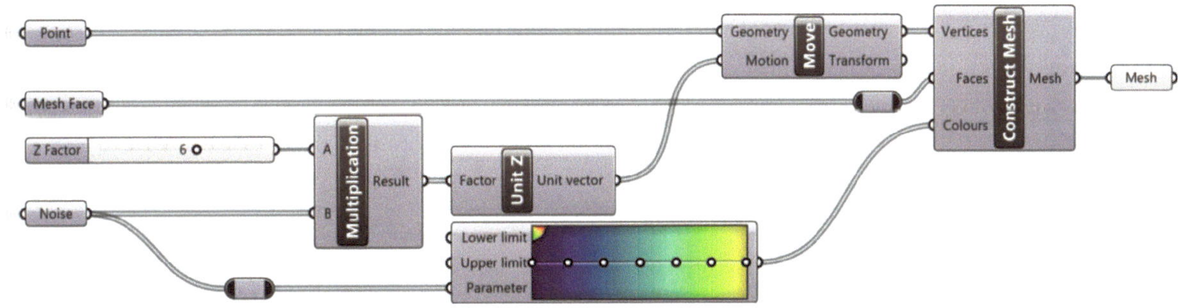

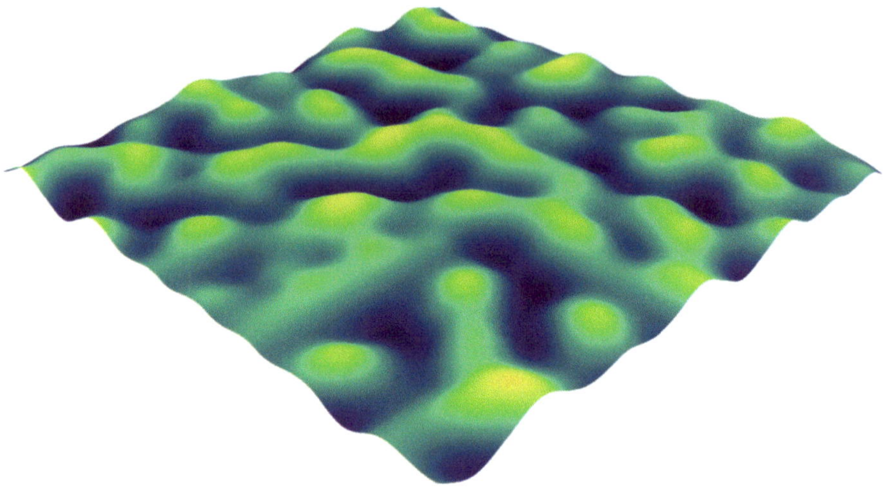

13.2.2 Cellular Noise

To generate a 3D mesh from cellular Worley noise, use the `Procedural Cellular` component from Bitmap+ instead. Use number sliders to set a random seed, the width and height of the grid, the depth, and frequency for the cellular noise. Set the boundary mode to `Euclidean`, `Manhattan`, or `Natural`. Set the interpolation mode to `Value`, `Distance2`, `Dist2Add`, `Dist2Sub`, `Dist2Mul`, or `Dist2Div`.

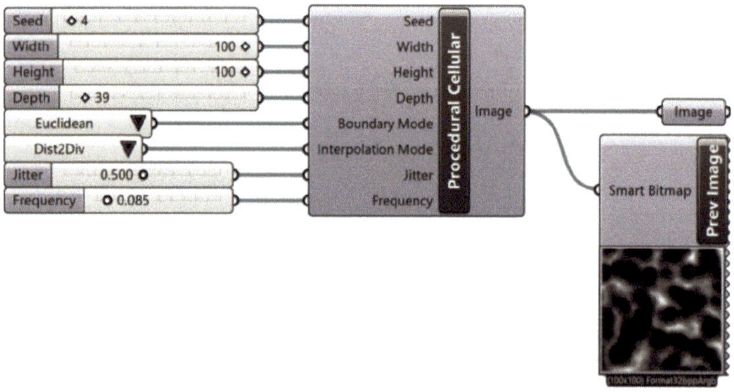

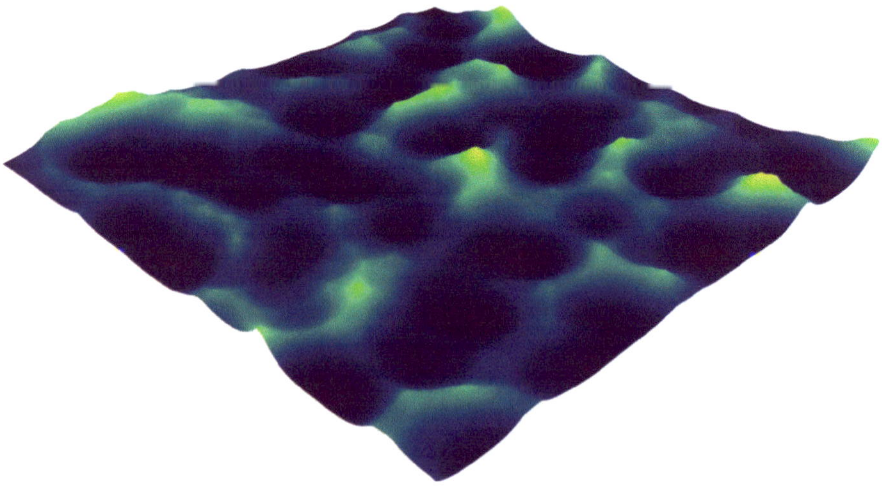

13.3 Noise in Python

There is a Python module – pyfastnoisesimd [2] – for the FastNoise library. This module can generate different types of noise such as white, value, Perlin, simplex, cellular, and cubic noise. It can generate fractal versions of value, Perlin, simplex, and cubic noise using fractional Brownian motion (fBm), billow, or ridged multifractal algorithms. The noise can be warped using normalization, gradient, or fractal gradient algorithms.

13.3.1 Fractal Perlin Noise

Generate fractal Perlin noise with domain warping using the pyfastnoisesimd module.

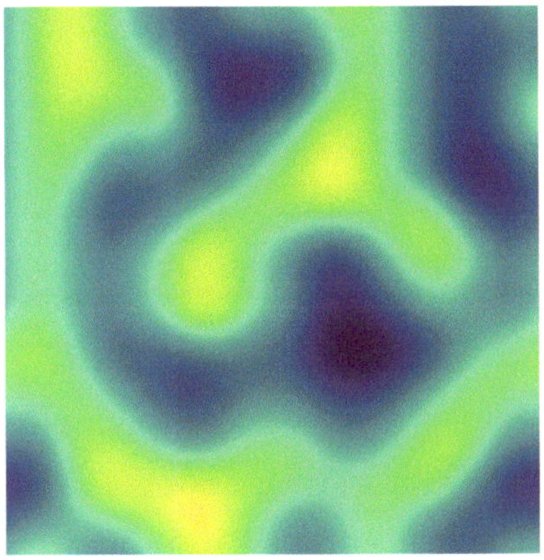

Perlin

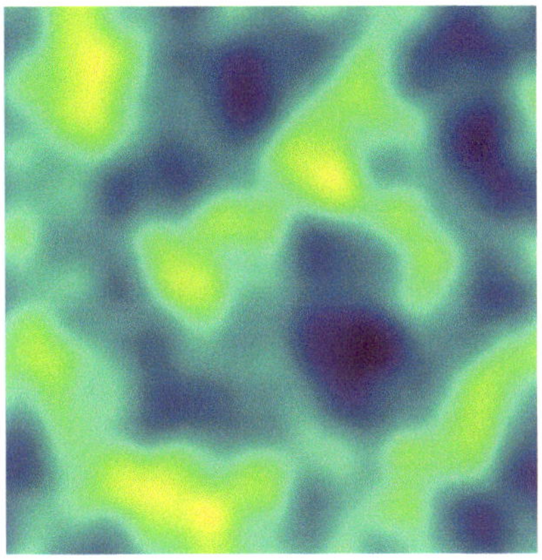

Perlin fBm

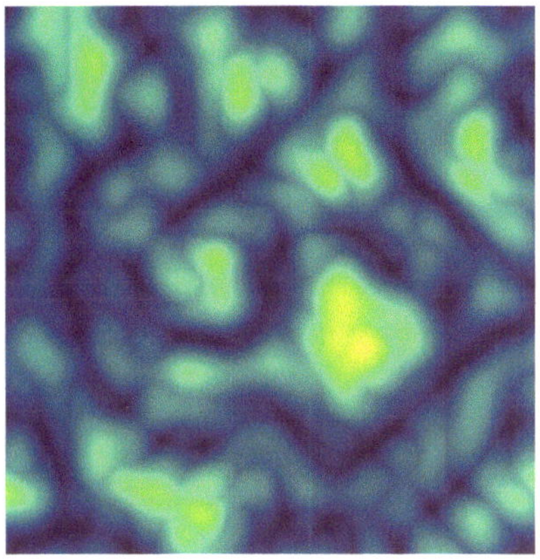

Perlin billow

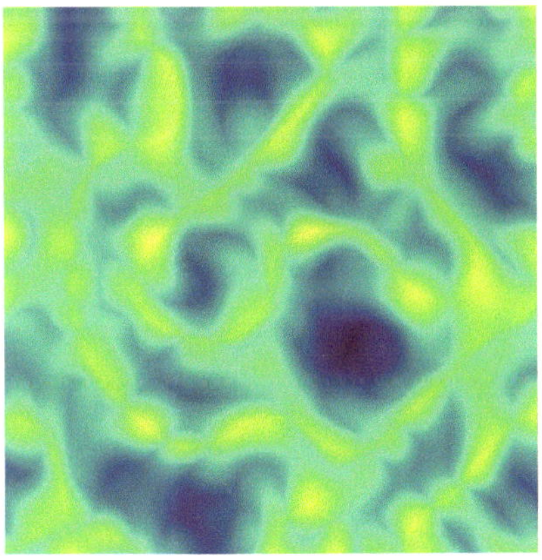

Perlin multifractal

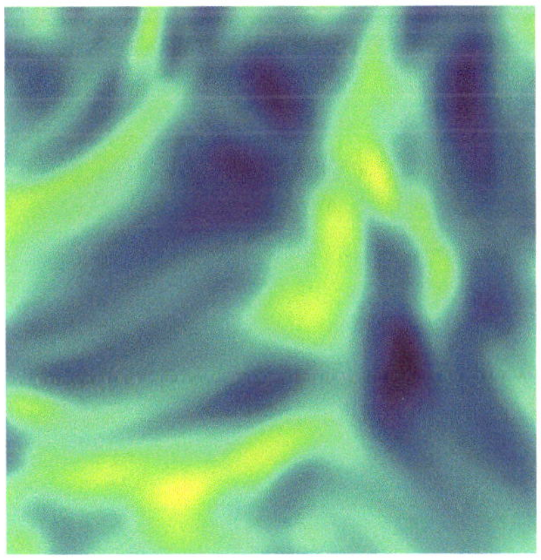

Perlin fBm gradient

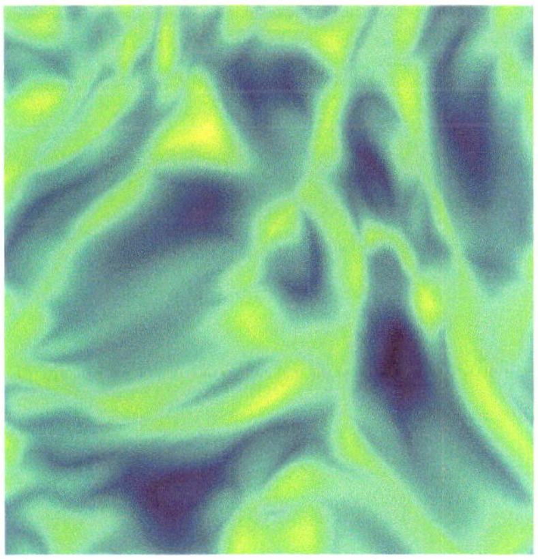

Perlin multifractal gradient

Perlin Noise Start by generating Perlin noise. First import the module pyfastnoisesimd for generating procedural noise, import the Seaborn library for plotting, and set a Seaborn theme. With the pyfastnoisesimd module the dimensions of the grid of noise must be divisible by the number of threads for parallel processing. Start by setting the number of threads to the number of cores your computer has with the `pyfastnoisesimd.num_virtual_cores` function. Then set the x- and y-dimensions of the grid as multiples of number of threads and add these dimensions to a list. Instantiate a noise object from the `pyfastnoisesimd.pyfastnoisesimd.Noise` class. Setting the attributes of this object will determine the type of noise, its frequency, random seed, and other properties. Set the `pyfastnoisesimd.Noise.noiseType` attribute to `Perlin`, set the `pyfastnoisesimd.Noise.frequency` to a floating point number, and optionally set the `pyfastnoisesimd.Noise.seed` to an integer for replicable results. To plot the noise in a grid, set the object's `pyfastnoisesimd.Noise.genAsGrid` attribute to the list of x- and y-dimensions. Finally plot the grid as a heatmap with `seaborn.heatmap`.

Code: `perlin-noise.py`

```
# import libraries
import pyfastnoisesimd as fns
import seaborn as sns

# set theme
sns.set_theme(style="white")

# set threads
threads = fns.num_virtual_cores()

# set dimensions
x = threads * 100
y = threads * 100
shape = [x, y]

# generate perlin noise
perlin = fns.Noise()
perlin.noiseType = fns.NoiseType.Perlin
perlin.frequency = 0.02
perlin.seed = 1

# plot perlin noise
noise = perlin.genAsGrid(shape=shape)

# generate heatmap
sns.heatmap(noise, cmap='viridis')
```

Fractal Noise To generate fractal noise, set the `pyfastnoisesimd.Noise.noiseType` attribute to a fractal variant – `ValueFractal`, `PerlinFractal`, `SimplexFractal`, or `CubicFractal`. Then set the fractal properties of the noise. First set the type with `pyfastnoisesimd.FractalType` to `FBM`, `Billow`, or `Rigid`. Then set the octaves, lacunarity, and gain.

Code: `perlin-fractal-noise.py`

```
# generate Perlin noise
perlin = fns.Noise()
perlin.noiseType = fns.NoiseType.PerlinFractal
perlin.frequency = 0.005
perlin.seed = 1

# set fractal properties
perlin.fractal.fractalType = fns.FractalType.FBM
perlin.fractal.octaves = 2
perlin.fractal.lacunarity = 2.1
perlin.fractal.gain = 0.45
```

Domain Warping To warp the noise, set the `pyfastnoisesimd.PerturbType` attribute to `Gradient` or `GradientFractal`.

Code: `perlin-fractal-gradient-noise.py`

```
# set perturbation properties
perlin.perturb.perturbType = fns.PerturbType.Gradient
perlin.perturb.amp = 1
perlin.perturb.frequency = 0.5
```

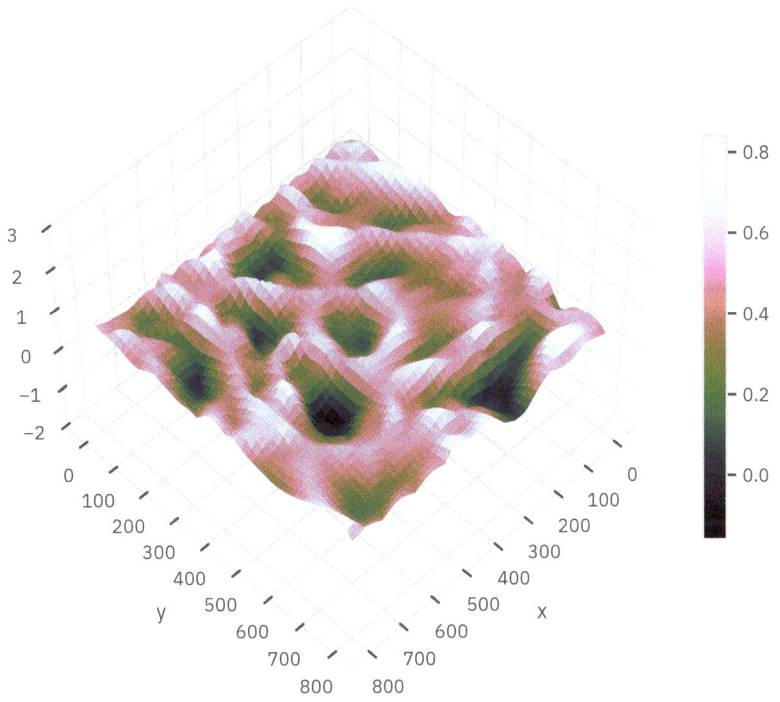

13.3.2 Cellular Noise

Generate cellular noise with domain warping using the pyfastnoisesimd module.

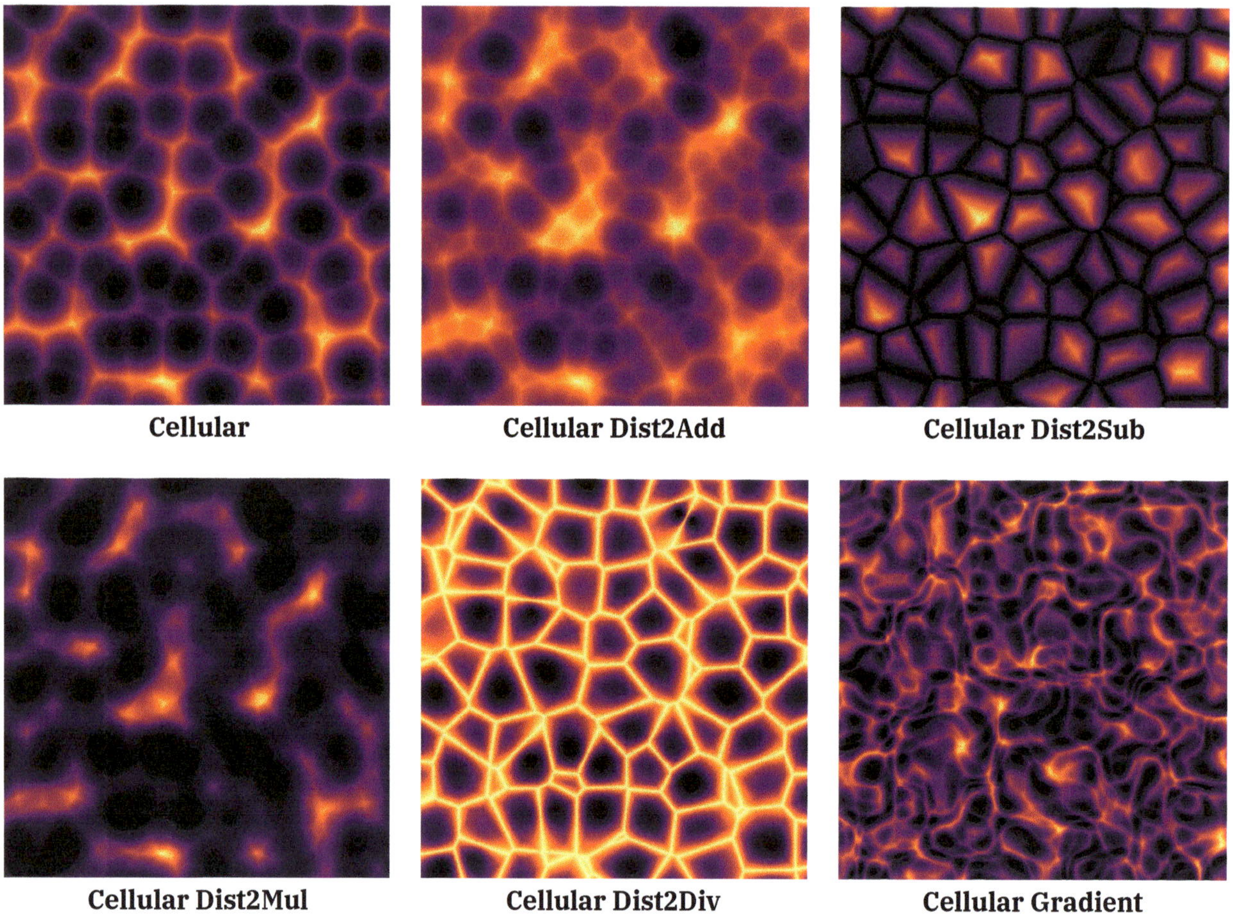

Cellular Noise To generate cellular noise, set the `pyfastnoisesimd.Noise.noiseType` attribute to `Cellular` and then set the cell properties. Set the `pyfastnoisesimd.CellularReturnType` attribute to one of the following options. `CellValue` calculates the value of the closest point; it generates a Voronoi diagram. `Distance` calculates the first-order distance, i.e. the distance to the closest point. `Distance2` calculates the second-order distance, i.e. the distance to the second closest point. `Distance2Add` calculates the sum of the first- and second-order distances. `Distance2Sub` calculates the difference of the first- and second-order distances. `Distance2Mul` calculates the product of the first- and second-order distances. `Distance2Div` calculates the quotient of the first- and second-order distances. Then set the `pyfastnoisesimd.CellularDistanceFunction` attribute to `Euclidean`, `Manhattan`, or `Natural` and set `jitter` to a floating point number.

Code: `cellular-noise.py`

```
# import libraries
import pyfastnoisesimd as fns
import seaborn as sns

# set theme
sns.set_theme(style="white")
```

```python
# set threads
threads = fns.num_virtual_cores()

# set dimensions
x = threads * 100
y = threads * 100
shape = [x, y]

# generate cellular noise
cellular = fns.Noise()
cellular.seed = 1
cellular.noiseType = fns.NoiseType.Cellular
cellular.frequency = 0.01

# set cellular properties
cellular.cell.returnType = fns.CellularReturnType.Distance
cellular.cell.distanceFunc = fns.CellularDistanceFunction.Euclidean
cellular.cell.jitter = 0.45

# plots cellular noise
noise = cellular.genAsGrid(shape=shape)

# generate heatmap
sns.heatmap(
    noise,
    cmap='inferno',
    annot=False,
    cbar=False,
    yticklabels=False,
    xticklabels=False,
    square=True
    )
```

Domain Warping To warp the cellular noise, set the `pyfastnoisesimd.PerturbType` attribute to `Gradient`.

Code: `cellular-gradient-noise.py`

```python
# set perturbation properties
cellular.perturb.perturbType = fns.PerturbType.Gradient
cellular.perturb.amp = 1.0
cellular.perturb.frequency = 0.5
```

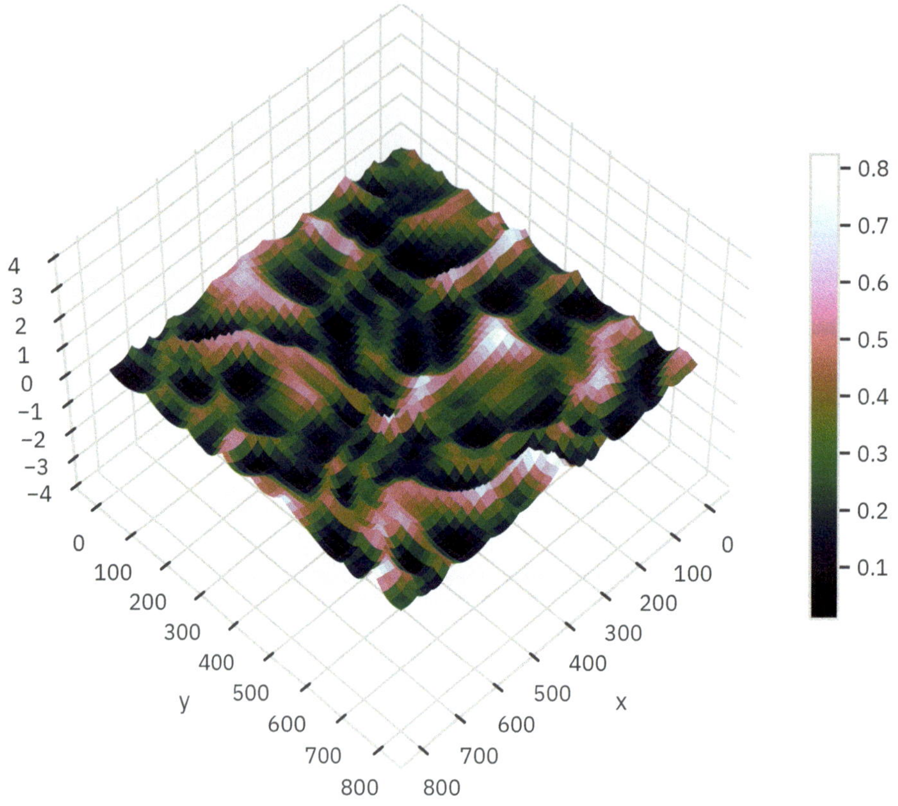

References

[1] [Software] David Mans, *Bitmap+* version 1.6.2, 2023. lic: MIT License. url: https://interopxyz.gitbook.io/bitmap-plus/, vcs: https://github.com/interopxyz/BitmapPlus.

[2] [Software] Robert McLeod, *PyFastNoiseSIMD* version 0.4.2, 2021. lic: BSD 3-Clause License. url: http://pyfastnoisesimd.readthedocs.io/, vcs: https://github.com/robbmcleod/pyfastnoisesimd.

[3] [Software] Jordan Peck, *FastNoise Lite* version 1.0.3, 2021. lic: MIT License. vcs: https://github.com/Auburn/FastNoiseLite.

[4] Daniela Mitterberger and Tiziano Derme. “Digital soil: Robotically 3D-printed granular biocomposites”. In: *International Journal of Architectural Computing* 18.2 (2020), pages 194–211. doi: 10.1177/1478077120924996.

[5] Gramazio Kohler Research. *Augmented Bricklaying*. 2019. url: https://gramaziokohler.arch.ethz.ch/web/e/projekte/371.html (visited on 07/22/2022).

[6] Inigo Quilez. *fBm*. 2019. url: https://iquilezles.org/articles/fbm/ (visited on 07/24/2022).

[7] Laura U. Marks. “A Noisy Brush with the Infinite: Noise in Enfolding-Unfolding Aesthetics”. In: *The Oxford Handbook of Sound and Image in Digital Media*. Oxford University Press, Dec. 2013. isbn: 9780199757640. doi: 10.1093/oxfordhb/9780199757640.013.001.

[8] Andrew Kudless. “Bodies in Formation: The material evolution of flexible formworks”. In: *Proceedings of the 31st Annual Conference of the Association for Computer Aided Design in Architecture*. ACADIA 11: Integration through Computation. Banff, Canada, 2011, pages 98–105. isbn: 978-1613645956. doi: 10.52842/conf.acadia.2011.098.

[9] [Software] Giulio Piacentino, *4D Noise* version 1, 2011. McNeel Europe. vcs: https://www.food4rhino.com/en/app/4d-noise.

[10] matsys. *P_Wall(2009)*. 2009. url: https://www.matsys.design/p_wall-2009 (visited on 07/22/2022).

[11] Ruth Jarman and Joe Gerhardt. *Brilliant Noise*. Semiconductor, 2006. url: https://semiconductorfilms.com/art/brilliant-noise/ (visited on 07/04/2023).

[12] Aden Evens. *Sound Ideas: Music, Machines, And Experience*. Volume 27. Theory out of Bounds. Minneapolis: University of Minnesota Press, 2005. isbn: 9781452907307.

[13] Forest Kenton Musgrave. "Fractal Terrains and Fractal Planets". In: *The Elements of Nature: Interactive and Realistic Techniques*. Edited by Oliver Deusen, David S. Ebert, Ron Fedkiw, Forest Kenton Musgrave, Przemyslaw Prusinkiewicz, Doug Roble, Jos Stam, and Jerry Tessendorf. SIGGRAPH '04. Los Angeles, CA: Association for Computing Machinery, 2004, pages 7.1–7.45. isbn: 9781450378017. doi: 10.1145/1103900.1103932.

[14] Forest Kenton Musgrave. "Fractal Solid Textures: Some Examples". In: *Texturing and Modeling*. Edited by David S. Ebert, Forest Kenton Musgrave, Darwyn Peachey, Ken Perlin, Steven Worley, William R. Mark, and John C. Hart. Third Edition. The Morgan Kaufmann Series in Computer Graphics. San Francisco: Morgan Kaufmann, 2003, pages 447–487. isbn: 978-1-55860-848-1. doi: 10.1016/B978-155860848-1/50044-9.

[15] Forest Kenton Musgrave. "Procedural Fractal Terrains". In: *Texturing and Modeling*. Edited by David S. Ebert, Forest Kenton Musgrave, Darwyn Peachey, Ken Perlin, Steven Worley, William R. Mark, and John C. Hart. Third Edition. The Morgan Kaufmann Series in Computer Graphics. San Francisco: Morgan Kaufmann, 2003, pages 489–506. isbn: 978-1-55860-848-1. doi: 10.1016/B978-155860848-1/50045-0.

[16] Ken Perlin. "Improving Noise". In: *ACM Transaction on Graphics* 21.3 (July 2002), pages 681–682. issn: 0730-0301. doi: 10.1145/566654.566636.

[17] Inigo Quilez. *Domain Warping*. 2002. url: https://iquilezles.org/articles/warp/ (visited on 07/24/2022).

[18] Steven Worley. "A Cellular Texture Basis Function". In: *Proceedings of the 23rd Annual Conference on Computer Graphics and Interactive Techniques*. SIGGRAPH '96. New York, NY, USA: Association for Computing Machinery, 1996, pages 291–294. isbn: 0897917464. doi: 10.1145/237170.237267.

[19] Forest Kenton Musgrave. "Methods for Realistic Landscape Imaging". PhD thesis. USA: Yale University, 1993.

[20] Ken Perlin. "An Image Synthesizer". In: *ACM SIGGRAPH Computer Graphics* 19.3 (July 1985), pages 287–296. issn: 0097-8930. doi: 10.1145/325165.325247.

14 Biome

14.1 Procedural Landscape

Generate a low-poly landscape with rugged terrain and woodlands using cellular noise. Start by modeling a very simple conifer from geometric primitives. Then generate a terrain mesh from cellular noise and plant the trees on the mesh, thinning the trees at high elevation. Scale the trees relative to their elevation and apply a color gradient based on elevation.

Plugin: `Bitmap+`

Code: `biome.gh`

14.2 Tree

Model a conifer in a low-poly style. Create a cylinder for the trunk and a series of cones for the canopy. Model this in Rhino or use the Grasshopper definition `lowpoly-tree.gh`.

Code: `lowpoly-tree.gh`

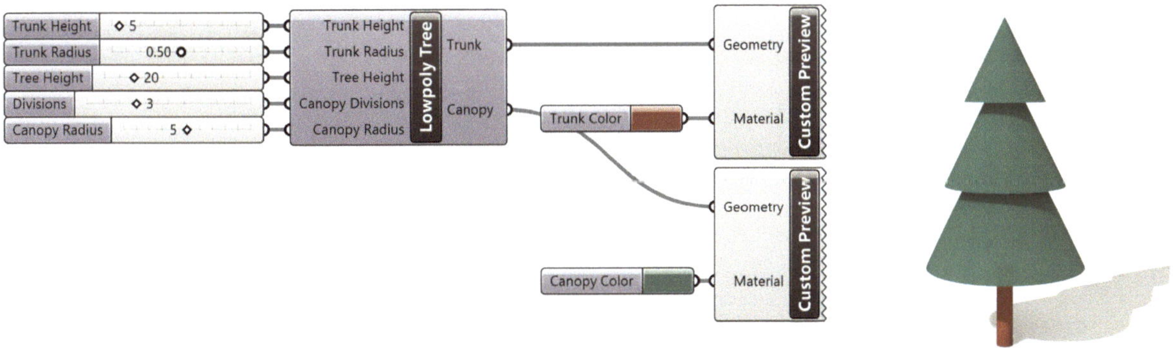

14.3 Terrain

Generate a terrain mesh from cellular noise.

Code: `fast-noise.gh`

Cellular Noise First generate a cellular noise map with `Bitmap+`'s `Procedural Cellular` component.

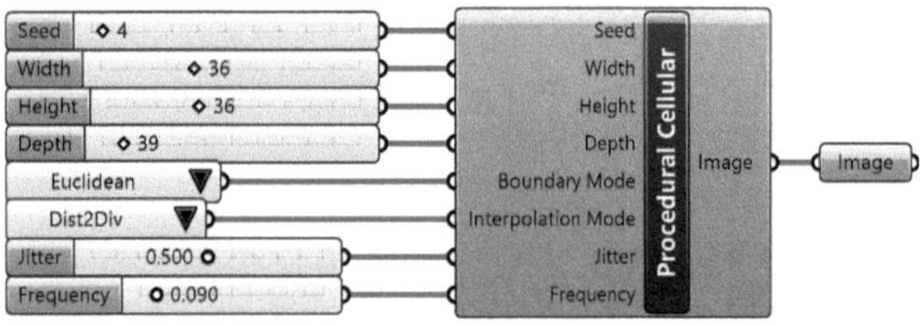

Grid Then create a mesh grid with the same number and size of cells as the image. Find the boundary, width, and height of the image with a `Bitmap Properties` component. Use these properties to construct a mesh grid with a `Mesh plane` component and then extract the vertices and faces of the mesh with a `Deconstruct Mesh` component.

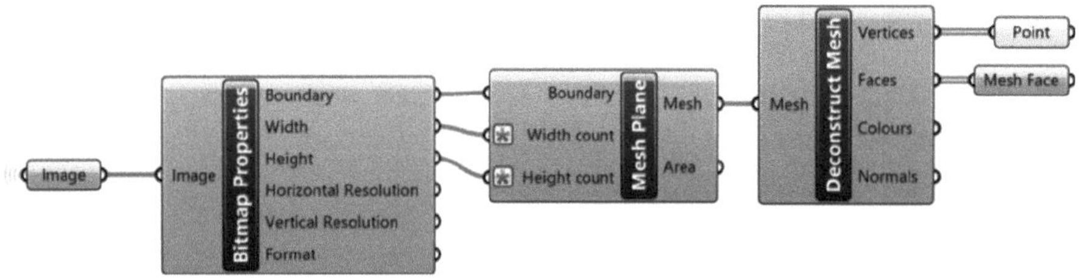

Noise Extract elevation values from the intensity of the image by deconstructing the image into colors with the `Deconstruct Bitmap` component and then splitting the colors with the `Split AHSV` component.

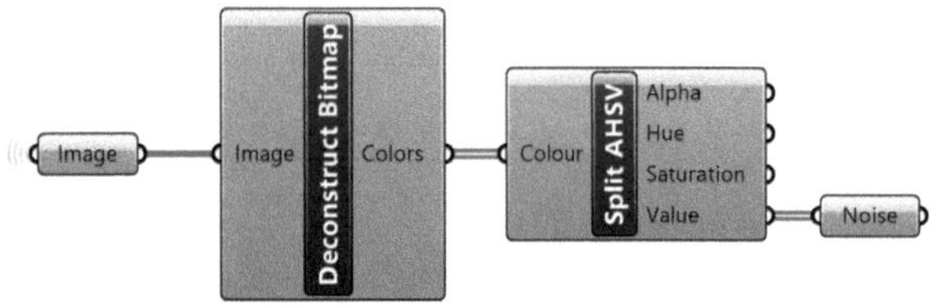

Mesh Generate a terrain mesh by moving the vertices of the mesh grid vertically based on the intensity of the image. First, multiply the intensity values by a scale factor. Then define the vertical motion by plugging the scaled intensity values into a `Unit Z` component. Move the mesh vertices along this translation vector with a `Move` component. Use these translated vertices, the original mesh faces, and a topographic color gradient to reconstruct the mesh with a `Construct Mesh` component.

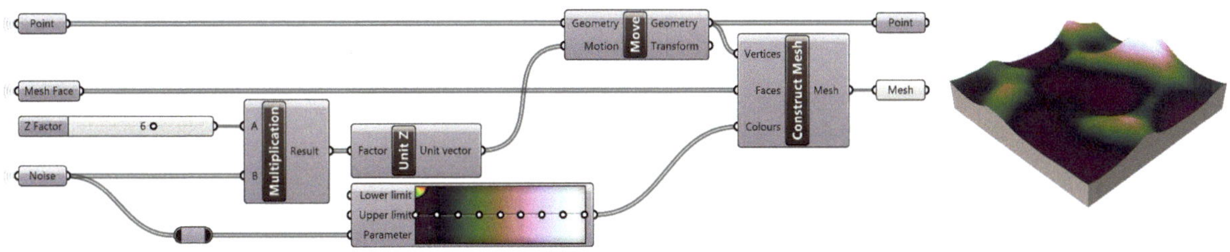

14.4 Ecosystem

Create an ecosystem with larger trees at lower elevations. For a simple approach, plant a tree on each vertex of the terrain mesh. Cull trees based on the intensity of the noise map, i.e. the elevation gradient, so that there are no trees at high elevation. Then randomly remove and jitter the trees so that they are more naturally distributed. Assign a color gradient to the trees based on elevation. For an alternative approach using a decay function try `ecosytem.gh`.

Code: `biome.gh`

Gradient Create a planting location at each vertex of the terrain mesh and then cull locations above a given elevation. Test which intensity values from the noise map are smaller than a given threshold with a `Smaller Than` component. Set the threshold with a number slider. Use the resulting list of Boolean truth values as a culling pattern. Plug the lists of vertices and noise values into `Cull Pattern` components to filter out higher values.

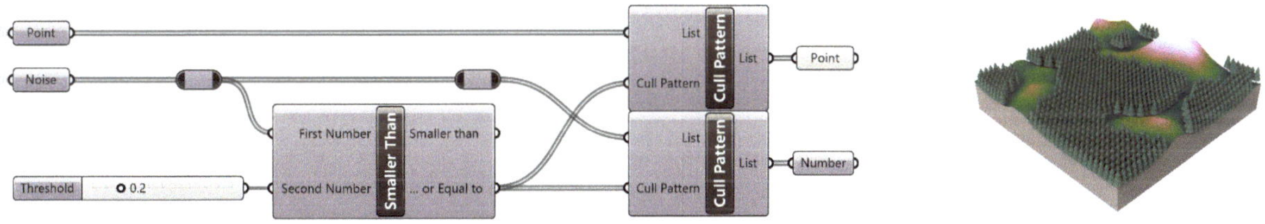

Decimation Decimate the culled lists of vertices, i.e. planting locations, and noise values. This will begin to disrupt the order of the grid, but will not be enough to form a natural distribution. Find the number of remaining planting locations with a `List Length` component. Multiply this by a reduction factor using a `Multiplication` component. Then decimate both lists with `Random Reduce` components using this scaled reduction factor.

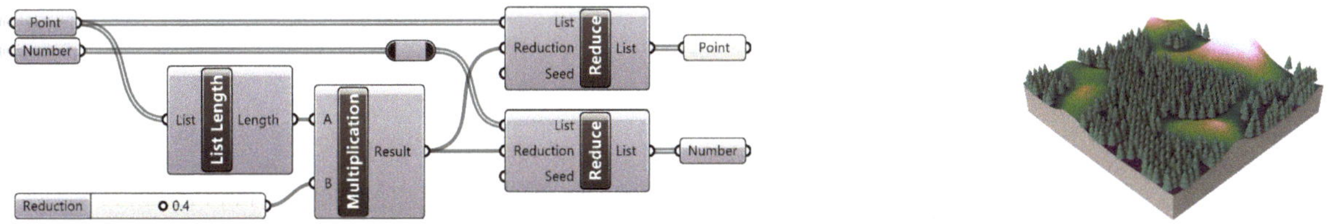

Jitter Jitter the planting locations for a more natural distribution of trees. To move each location a short distance in a random direction, first deconstruct these points into their x-, y-, z-components with a `Deconstruct` component. Then find the number of remaining locations with a `List Length` component and generate a list of random values of that length with a `Random` component. Add these random values to the x- and y-components of the points with `Addition` components and the reconstruct the points with `construct Point` components.

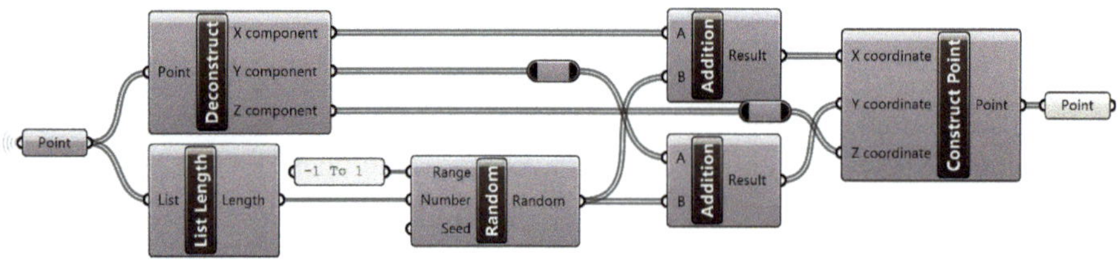

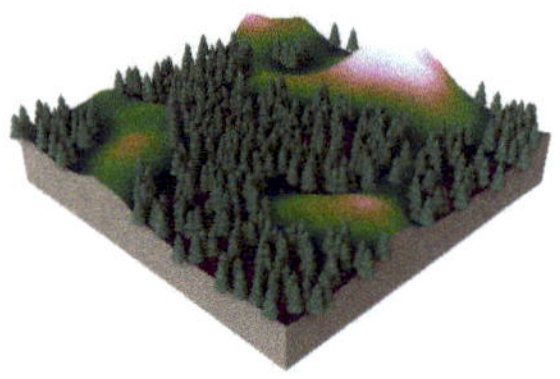

Snapping Jittering the planting locations may cause some of these points to fall off the terrain. Use the Kangaroo physics engine to snap these points back onto the terrain mesh. With Kangaroo, the points will be treated as particles in a particle system. First set snapping these particles onto the terrain mesh as a goal for the physics engine with an `On Mesh` component. Then run the physics simulation with a `Zombie Solver` component.

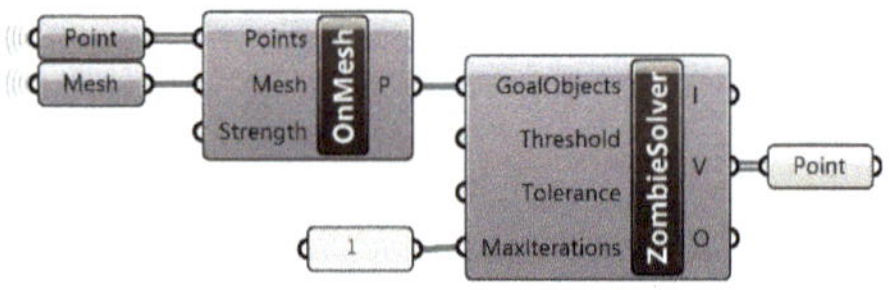

Transform Distribute low-polygon tree models to each planting location and then scale each model based on its elevation. First, move a tree trunk and its canopy to each point with `Move` components. Then scale the trunks and canopies with `Scale` components. Set the center of scaling to each point. Set the scale factor to a remapped list of noise. To do this, remap the list of noise values with a `Remap Numbers` component. Find the source domain with a `Bounds` component and set the target domain from `1 To 0.5`.

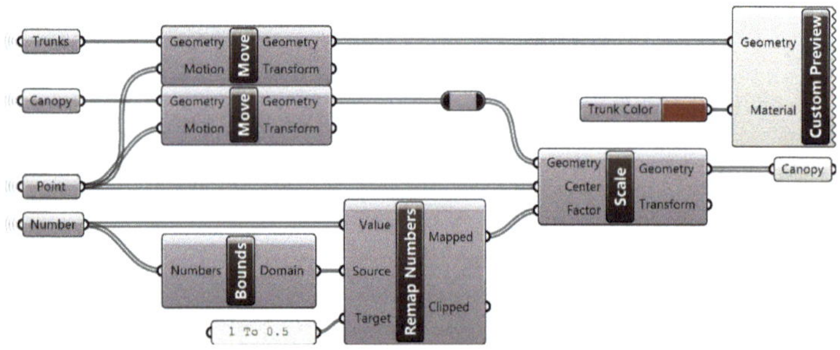

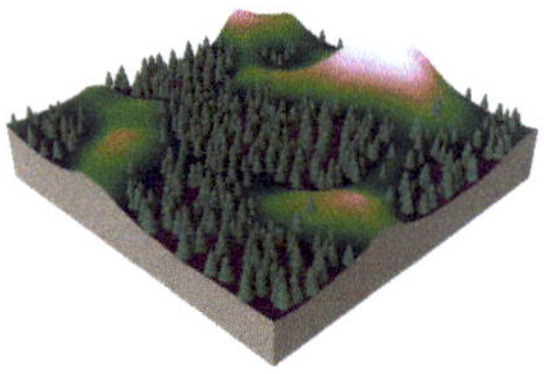

Color Assign colors to the tree canopy based on the elevation gradient. Visualize the canopy with a `Custom Preview` and define its material with a `Gradient` component. Set the parameters for the gradient to a remapped list of noise.

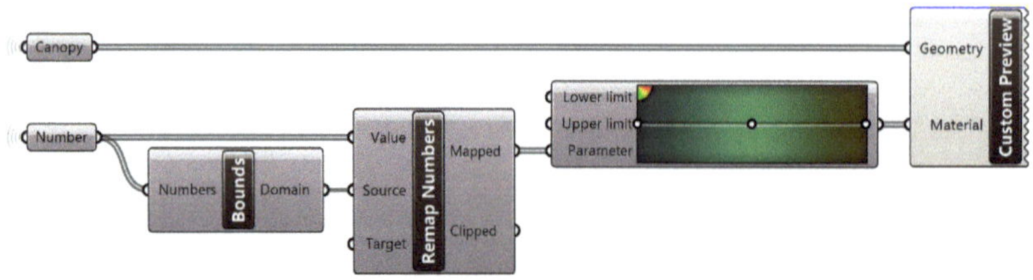

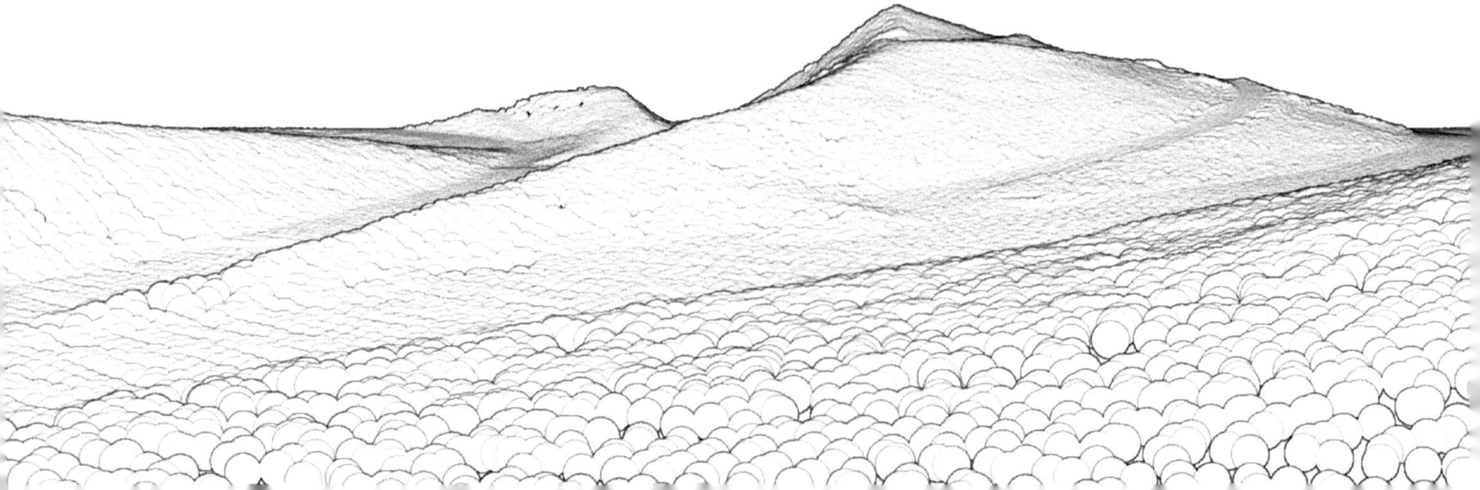

15 Terrain

Terrain is a fundamental medium of landscape architecture. Its form, from the ground underfoot to the peaks in the distance, shapes our experience of a landscape. The steepness of a slope is felt in our legs as we climb or descend. The shape of the terrain opens or closes our view, framing the landscape. Its form directs processes such as the flow of water and sediment. Landscape architects model terrain, analyze its form and the processes it affects, and regrade it – giving it new form. These designed landforms can be powerful aesthetic statements. Famous earthworks include Jean-Charles Adolphe Alphand's Parc des Buttes Chaumont, Michael Heizer's Effigy Tumuli, Charles Jencks' Landform Ueda [45], and Maya Lin's Eleven Minute Line [42]. This chapter introduces data structures and methods for modeling terrain. Our study area will be the Hills of Governor's Island, designed by West 8 [32]. The terrain dataset for this chapter is from a 2017 airborne lidar survey of New York City at 1-foot resolution [29].

15.0.1 Data Structures

Traditionally landscape architects have represented terrain as spot elevations and contour curves. Computationally, however, terrain can also be represented as arrays, point clouds, rasters, and meshes (Figure 15.1). While a point cloud is a unstructured collection of spatial coordinates, a raster is a regular grid of data. The grid itself demarcates space, while its cells store data such as elevation, slope, or color. A mesh is a network of polygons that defines a shape. Each polygon is composed of a face, bound by edges, connecting its vertices. Each representation has its uses. While arrays and rasters enable easy numerical calculations, meshes are used for 3D rendering and digital fabrication. Rasters are used for quantitative analyses of terrain including slope, aspect, curvature, hydrology, viewsheds, and landform classification [26].

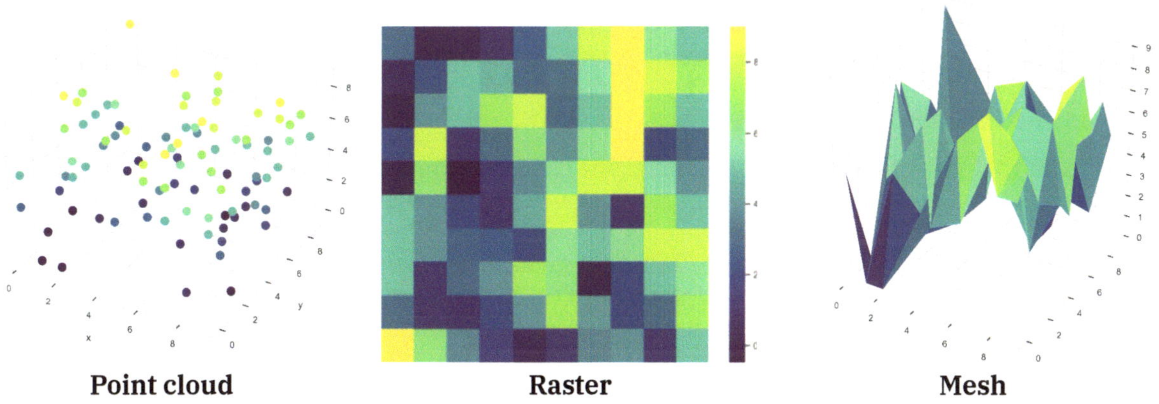

Figure 15.1: Data structures

Point Clouds As covered in chapter 7, dense, high-resolution point clouds can be collected with airborne lidar, drone lidar, or drone photogrammetry. While airborne lidar captures everything hit by laser pulses including ground, trees and buildings, it can be filtered into different classes, either manually or automatically (Figure 15.2). Standard lidar classes include **1** unclassified, **2** ground, **3** low vegetation, **4** medium vegetation, **5** high vegetation, **6** buildings, and **9** water [17]. Algorithms for automatically classifying bare ground points include the Multiscale Curvature Classification (MCC) algorithm [39]. Rasters can be generated from point clouds by binning or interpolation, while meshes can be generated with triangulation.

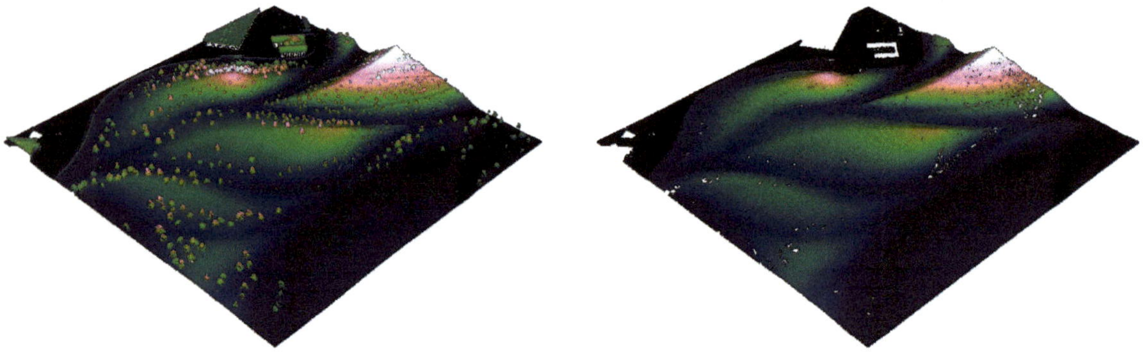

Lidar: All classes **Lidar:** Ground class

Figure 15.2: Lidar classes

Raster Terrain rasters represent elevation z as a function of x and y coordinates on a grid.

$$z = f(x, y) \quad \textbf{Equation } 15.1$$

A terrain raster is a matrix of cells where the grid maps x- and y-space and the cells are assigned elevation values [37]. The resolution of a terrain raster is the size of the grid cells in angular or linear units such as degrees, meters, or feet (Figure 15.3). Rasters that represent the elevation of bare ground are called digital elevation models, while rasters that represent the height of everything on and above the ground are called digital surface models. Rasters representing the height of canopy above the ground are called canopy height models. These can be calculated by subtracting the digital elevation model from the digital surface model. Digital elevation models can be generated from the ground class of classified lidar point clouds. The rasters used in this book are GeoTIFFs (`.tif`) – a common, georeferenced raster format for spatial data. GeoTIFF is a public domain standard published by the Open Geospatial Consortium [18].

Terrain rasters can be generated from point clouds by binning or interpolation. In binning, the points falling in each cell – i.e. bin – of the raster grid are aggregated using statistical methods. Mean statistics are used when binning digital elevation models, while maximum statistics are used for binning canopy height models. Binning is fast and efficient, but large distances between irregularly distributed points can result in gaps in the raster. Interpolation fits a continuous surface through a set of points, filling gaps in the data. Rasters can be interpolated from points using local neighborhood, geostatistical, and variational approaches. Methods include triangulated irregular network, inverse distance weighted, natural neighbor, kriging, and spline interpolation [46]. While in binning raster resolution should be greater than point spacing so that every grid cell has at least one point, interpolation can be used for supersampling. Methods such as regularized spline with tension interpolation can generate smooth topographic surfaces that are appropriate for calculating topographic parameters such as slope and curvature and modeling processes such as the flow of water and sediment [41, 36].

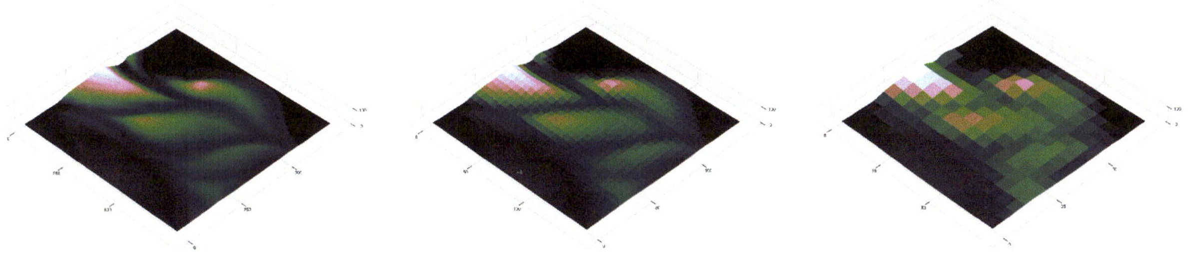

Resolution: 1 ft **Resolution:** 5 ft **Resolution:** 10 ft

Figure 15.3: Raster resolution

Mesh Terrain is often represented as a triangular mesh. A mesh can be generated from a set of points – such as a lidar point cloud – by Delaunay triangulation. In Delaunay triangulation, the circumcircle of each triangle must be empty, i.e. no other points should fall within the circle intersecting the vertices of a triangle [48, 47]. This maximizes the minimum angle of the triangles yielding more aesthetic meshes. The centers of the circumcircles form the vertices of a cellular graph called a Voronoi tessellation (Figure 15.4). Learn more about Voronoi diagrams in chapter 6.

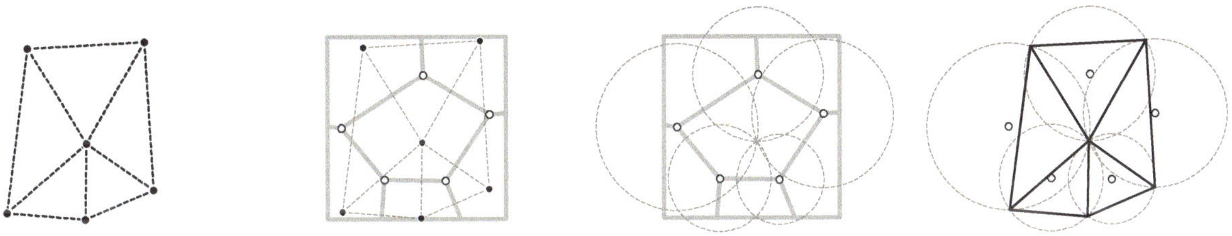

Delaunay mesh **Voronoi tessellation** **Voronoi vertices** **Circumcircles**

Figure 15.4: Delaunay triangulation

15.0.2 Data Sources

Topographic data is widely available as rasters and increasingly as point clouds. Open Topography has an extensive collection of topographic and bathymetric data including lidar point clouds and derived digital elevation models [38]. Both Open Topography and Google Earth Engine [44, 27] distribute regional and global topographic datasets (Table 1).

Table 1: Topographic datasets

Agency	Dataset	Extent	Resolution	Source
USGS	3D Elevation Program (3DEP)	US	30m | 10m | 3.4m	[31]
ESA	Copernicus European DEM (EEA-10)	Europe	10m	[19]
NASA	NASADEM	Global	30m	[16]
JAXA	ALOS Global Digital Surface Model (AW3D30)	Global	30m	[33, 34]
ESA	Copernicus Global DEM (GLO-90/-30)	Global	90m | 30m	[19]
NASA	Shuttle Radar Topography Mission (SRTM)	Global	90m | 30m	[43, 40]
UTokyo	MERIT DEM	Global	90m	[24, 30]

15.1 Grasshopper

In Rhino terrain can be modeled as a mesh for rendering, computing contours, 3D printing, and CNC machining. Plugins for terrain modeling in Rhino include RhinoTerrain [9], which can import both rasters and lidar point clouds and efficiently model very large triangulated meshes. Plugins for terrain modeling and grading in Grasshopper include Docofossor [21, 20], Land Kit [7], Groundhog [11], and Bison [23]. This section covers how to triangulate a mesh from a point cloud using native Grasshopper components.

15.1.1 Terrain Mesh

Model the Hills of Governor's Island as a terrain mesh from a lidar point cloud. Start by opening the Rhino model `hills-of-governors-island.3dm` which contains a point cloud from airborne lidar. Use the Grasshopper definition `terrain-mesh.gh` as a reference.

Model: `hills-of-governors-island.3dm` **Code:** `terrain-mesh.gh`

Delaunay Triangulation Start Grasshopper and add a point parameter to the canvas. Right click on the point parameter and set it to the exploded point cloud in the Rhino scene. Plug the point parameter into a `Delaunay Mesh` component to triangulate a mesh between the points. To assign hypsometric tinting – i.e. an elevation color map – deconstruct the mesh, extract the z-component of its vertices, map a color gradient to the z-values, and then reconstruct the mesh with this new color data. First, plug the mesh into a `Deconstruct Mesh` component to extract its vertices and faces. Then plug the vertices into a `Deconstruct Point` component to extract their x-, y-, and z-components. Remap the list of z-components to a domain from 0 to 1. To do this, plug the z-component into `Bounds` to find its domain. Then plug the z-component into `Remap numbers` as the input value and plug the domain from `Bounds` in as the source. This will remap the elevation to a gradient from 0 to 1. Plug the mapped values into a `Gradient` component to assign colors. See appendix B for a guide to color maps. Finally reconstruct the mesh. Add a `Construct Mesh` component and plug in the deconstructed vertices, deconstructed faces, and color gradient as its inputs.

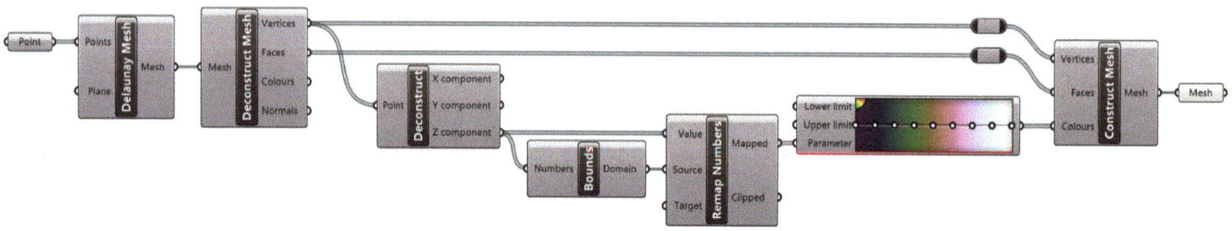

Contours Compute contours from the terrain mesh. Simply plug the mesh into a `Contour` component as the input shape. Set the distance, i.e. contour interval, with a number slider. By default the direction normal to the contours will be set to the z-axis. Optionally, add a `Custom Preview` component to better visualize the contours.

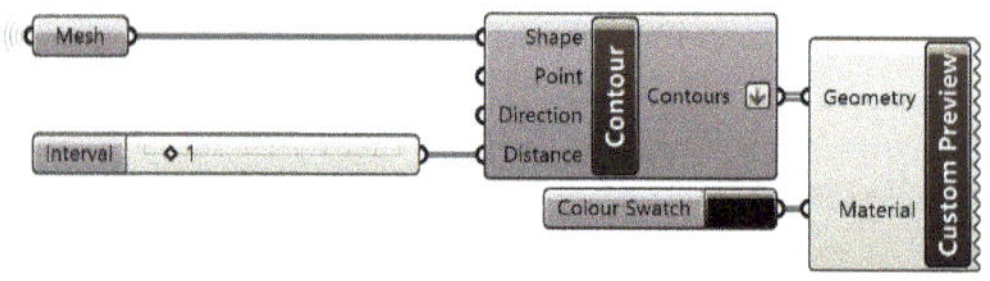

15.2 Python

There are Python libraries for importing, processing, and rendering terrain data. Rasters can easily be imported and plotted in Python with the `rasterio` library [35] which is built on GDAL – the geospatial data abstraction library [2]. Libraries for working with point clouds include `laspy` [1], `PyVista` [22, 6], and `Open3d` [25, 5]. For more sophisticated terrain modeling in Python, try `Landlab` or an open source geographic information system with a Python scripting library such as `GRASS GIS` [3] or `QGIS`, [8]. Landlab is a Python library for modeling Earth surface dynamics that supports rasters and hexagonal grids as well as networks and meshes [15, 28, 12]. This section covers importing, contouring, and plotting rasters with `rasterio`, importing point clouds with `laspy`, and rendering point clouds with `pyvista`.

15.2.1 Rasters

Use the `rasterio` library to import and plot a digital elevation model. Use the script `raster-import.py` as a guide. First import the library and its `rasterio.plot.show` function. Then set the full file path for the raster. See appendix C for a guide to setting file paths. Open the raster dataset with `rasterio.open`. This will return a dataset object with attributes including the number of bands, width and height of the array, bounds, transform, and coordinate reference system (CRS). Print the rasters' coordinate reference system by calling the object's `.crs` attribute inside a `print()` statement. For our example, this will print `EPSG:2263` for the North American Datum 1983, New York Long Island State Plane in US survey feet. Finally plot the dataset object using `rasterio.plot.show`. Set a color map such as `cubehelix` for the plot. The color gradient will represent the change in elevation. See appendix B for a guide to color maps. For a more complete example with styling and a legend, see the script `raster-legend.py`.

Dataset: `elevation.tif`

Code: `raster-import.py`

```
# import modules
import rasterio as rio
from rasterio.plot import show

# set path
dataset = r'elevation.tif'

# read raster
elevation = rio.open(dataset)
```

```
# print coordinate reference system
print(elevation.crs)

# plot raster
show(elevation, cmap='cubehelix')
```

```
EPSG:2263
```

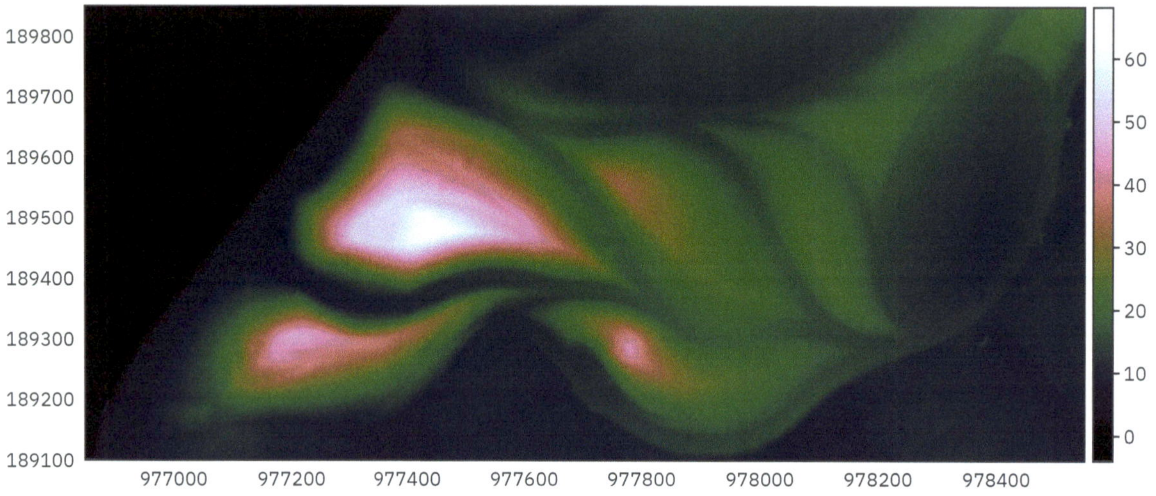

Contours Contours can easily be generated from a digital elevation model with `rasterio`. To plot contours simply call the function `rasterio.plot.show` with the `contour` parameter set to `True`. This example demonstrates how to plot contours over a digital elevation model with hypsometric tinting. Contours will be plotted at 3-foot intervals.

Code: `raster-contours.py`

Setting a custom contour interval for a Rasterio plot requires a list of contour levels. After opening the raster, read its elevation values as an array, find the elevation range, and then generate a list of 3 foot contour intervals with that range. First import `numpy`, `rasterio`, `rasterio.plot.show`, and `matplotlib.pyplot`. Then set the full file path to the raster. Open the raster as a dataset object with `rasterio.open`. Use the object's `read` method to return a NumPy array of elevation values `z`. Convert null values to zeros with `numpy.nan_to_num`. Set the interval to `2`, set the minimum contour level to `0`, and set the maximum contour level to highest elevation. Use the function `numpy.nanmax` to find the highest elevation. Then generate an array of contour levels with `numpy.arange`.

Create a figure with subplots to layer the contours on top of the digital elevation model. In Matplotlib, a figure is a complete sets of graphics that may contain a plot or multiple subplots and other elements such as data, labels, text, and a legend. An axis is a plot within a figure. Each axis contains its own data and other elements. Create a figure with a single subplot with `matplotlib.pyplot.subplots`. To layer the contours over the digital elevation model, plot them both on the same subplot. First plot the digital elevation model with `rasterio.plot.show` on subplot `ax`. Then plot the contours using `rasterio.plot.show` with source set to the digital elevation model, contours set to `True`, levels set to the array of contours levels, and the subplot set to `ax`.

```
# import modules
import numpy as np
import rasterio as rio
from rasterio.plot import show
from matplotlib import pyplot

# set path
dataset = r'elevation.tif'

# read raster
elevation = rio.open(dataset)
z = elevation.read()
z = np.nan_to_num(z)
minimum = 0
maximum = int(np.nanmax(z))
interval = 3
levels = np.arange(minimum, maximum, interval)

# plot raster
fig, ax = pyplot.subplots(figsize=(12, 12))
show(elevation, cmap='cubehelix', ax=ax)
show(
     elevation,
     colors='white',
     contour=True,
     levels=levels,
     linewidths=1,
     ax=ax
     )
```

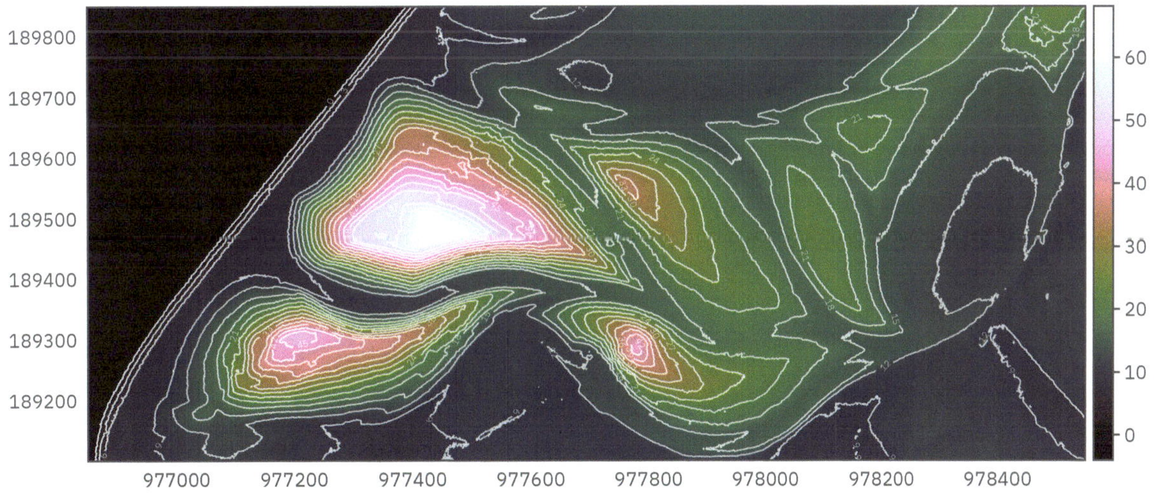

15.2.2 Point Clouds

With its robust support for different types of point clouds, Python is useful for processing topographic lidar data. In Python, `.laz` point clouds can be imported with the `laspy` [1] library and rendered using either the `PyVista` [22, 6] or `Open3D` [25, 5] library. This section covers reading, slicing, and rendering point clouds with hypsometric tinting or true color.

Slicing Use slicing to extract coordinates and colors from point clouds. Slicing is cutting apart a string, list, array, or so forth to extract part of it. It is an essential operation for working with arrays. Use slicing to access rows or columns in an array. The syntax for slicing a 1-dimensional array is `[start:stop:step]`, while the syntax for slicing a 2-dimensional array is: `[row, column]`.

Import a `.laz` file with the function `laspy.read`. This will return a `laspy.LasData` object with spatial and color data stored as attributes. Retrieve the coordinates of the point cloud as an array by calling `laspy.lasdata.LasData.xyz`. Try slicing the entire point cloud using `[:]` notation. This will select everything in every row.

Data: `all.laz`

```python
import laspy

# read lidar
dataset = r'all.laz'
las = laspy.read(dataset)
xyz = las.xyz

# print everything
print(xyz[:])
```

```
[[977475.55  189834.68   9.71]
 [977484.78  189849.50   9.84]
 [977484.24  189847.84   9.88]
 [977482.64  189843.03   9.94]
 ...
 [978221.06  189103.05   9.15]
 [978220.21  189100.76   9.05]
 [978224.03  189101.44   9.15]
 [978224.77  189103.47   9.28]]
```

Index the first row of the point cloud using `[0]` notation.

```python
# print first row
print(xyz[0])
```

```
[977475.55  189834.68   9.71]
```

Alternatively, slice the first row of the point cloud using `[:1]` notation.

```python
# print first row
print(xyz[:1])
```

```
[[977475.55  189834.68   9.71]]
```

Slice the first column with the x-values using `[:,0]` notation. The colon before the comma will select all rows. The index value of 0 after the comma will select the first column.

```python
# print first column
print(xyz[:,0])
```

```
[977475.55  977484.78  977484.24
...
978220.21  978224.03  978224.77]
```

Slice the second column with the y-values using `[:,1]` notation.

```python
# print second column
print(xyz[:,1])
```

```
[189834.68  189849.5   189847.84
...
189100.76  189101.44  189103.47]
```

Slice the third column with the z-values using `[:,2]` notation.

```
# print third column
print(xyz[:,2])
```

```
[9.71  9.84  9.88 ... 9.05  9.15  9.28]
```

Elevation Import a lidar point cloud and render it with a hypsometric color map. Read the `.laz` point cloud with Laspy and return the x-, y-, and z-values as an array. Plot the array with PyVista. Color the plot by setting a scalar field. Set `pyvista.plot`'s scalars to the z-values in the array and set the color map to `cubehelix`. Use slicing to extract the z-values. The slice `xyz[:, 2]` will return the third column of the array with the elevation values.

Data: `ground.laz`

Code: `topographic-lidar-hypsometric.py`

```
# import modules
import laspy
import pyvista as pv

# set path
dataset = r'ground.laz'

# read lidar
las = laspy.read(dataset)
xyz = las.xyz

# plot
pv.plot(
    xyz,
    scalars=xyz[:, 2],
    cmap='cubehelix',
    point_size=2,
    ambient=0.8,
    eye_dome_lighting=True
    )
```

Color Import a lidar point cloud and render it in color. First import Laspy, NumPy, and PyVista. Then set the path, read the `.laz` dataset, return the x-, y-, and z-coordinates, interweave the colors, and plot with `pyvista.plot`. Set the plot's scalars to the array of `rgb` colors.

Data: `all.laz`

Code: `topographic-lidar-rgb.py`

```python
# import modules
import laspy
import numpy as np
import pyvista as pv

# set path
dataset = r'all.laz'

# read lidar
las = laspy.read(dataset)
xyz = las.xyz

# plot with color
rgb = np.column_stack((
    las.red,
    las.green,
    las.blue
    ))
pv.plot(
    xyz,
    scalars=rgb,
    point_size=2,
    ambient=0.8,
    eye_dome_lighting=True,
    rgba=True,
    )
```

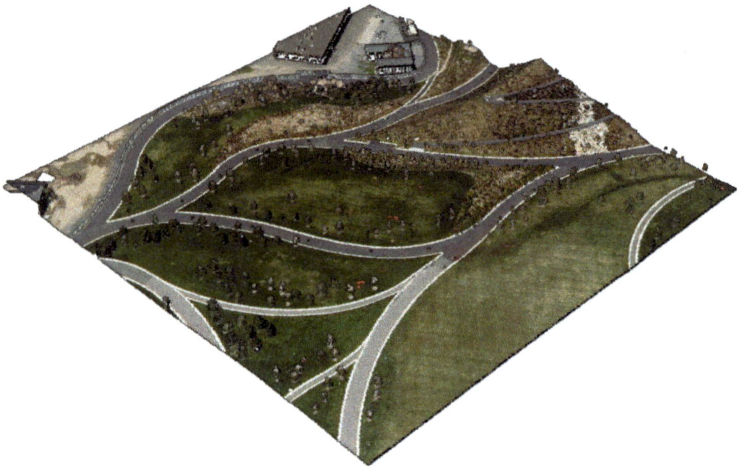

15.3 Rhino Scripting

Rhino does not have native support for `.las` point clouds or `.tif` georeferenced rasters, so either plugins or scripting are needed for importing many topographic datasets. While there are Rhino plugins such as `RhinoTerrain` and Grasshopper plugins such as `Heron` [10] that can import lidar and georeferenced raster data, scripting gives you more control over your data. This section covers how to import raster and lidar data into Rhino using the Python Script Editor. To import lidar, simply read the `.las` or `.laz` point cloud with `Laspy`, return the coordinates as an array, and add these to a Rhino point cloud. To import a georeferenced raster, first convert it into NumPy array with `Lidario` [13] and then add the array to a Rhino point cloud.

15.3.1 Importing Lidar

Use `Laspy` in Rhino' Script Editor to import a `.laz` color point cloud into Rhino. For a simpler example without color, see the script `import-laz.py`.

Data: `ground.laz`

Code: `import-color-laz.py`

Libraries Create a new Python script in Rhino's Script Editor. Set `lazrs`, `laspy`, and `numpy` as required packages. Then import `laspy`, `numpy`, and `rhinoscriptsyntax`.

Read Set the file path to the dataset `ground.laz`. Read this dataset with Laspy's `laspy.read` method. This will return a `laspy.LasData` object. The point cloud's coordinates can be accessed by calling the object's `.xyz` attribute. This will return an array of x-, y-, and z-values.

Decimate Point clouds captured with airborne lidar tend to have variations in point density. High variations in point density can result in visual artifacts in the point cloud and errors in derived products such as meshes and rasters. Artifacts and errors arising from point density include scan patterns, banding, swath overlaps, Moiré effects, and corduroy effects [4]. Decimating a point cloud removes redundant points, reducing computational costs and variations in point density. In this example, use count-based decimation to reduce the size of the array of coordinates. Use Python's slicing notation – `[start:stop:step]` – to remove every n^{th} point from the point cloud. For example the statement `xyz = las.xyz[::2]` will return every other set of coordinates, removing 50 percent of the points.

Colors The point cloud's colors can be accessed by calling the `laspy.LasData` object's `.red`, `.green`, and `.blue` attributes. Weave the arrays of red, green, blue color channels together with the function `numpy.column_stack`. Then decimate the array of colors and convert it into a list with the method `numpy.ndarray.tolist`.

Generate Generate a color point cloud in Rhino from the array of coordinates and the list of colors with RhinoScriptSyntax's `AddPointCloud` function. As it will be far from the origin of the scene, the georeferenced point cloud will fall outside the current view. To center the view on the point cloud, zoom all of the viewports to the extent of the geometry using Rhino's command line.

```
# requirements: lazrs, laspy, numpy
import laspy
import numpy as np
import rhinoscriptsyntax as rs

# set path
data = 'ground.laz'

# set variables
decimation = 2

# read lidar
las = laspy.read(data)

# decimate point cloud
xyz = las.xyz[::decimation]

# set colors
rgb = np.column_stack((las.red, las.green, las.blue))
rgb = rgb[::decimation].tolist()

# generate colored point cloud
rs.AddPointCloud(xyz, rgb)
```

```
Zoom
All
Extents
```

15.3.2 Exchanging Data

This section introduces a simple method for exchanging data between Python and Rhino. In Python you may create and work in different environments with different packages installed. For example, in this book, we have been using the `computational-design` Conda environment. You may also use Python scripting in other programs like Rhino with their own environments. If you generate data in one Python environment, how can you work with it in another? A simple method for exchanging data between environments is to save it as an array. Arrays are a fast, efficient, and widely used data structure that serve as the foundation for much of scientific computing [14]. In the first environment, convert the data to an array and save the array as a NumPy binary. Then in the second environment, import the NumPy binary as an array.

Arrays from Rasters This example covers how to convert a raster to an array and then save it as a NumPy binary using the `computational-design` Conda environment in Spyder.

Data: `elevation.tif`

Code: `geotiff-to-npy.py`

Start a new script in Spyder. Import `numpy` and `lidario`. Then set the path to the raster dataset `elevation.tif`. Import this georeferenced raster and convert it to a NumPy array with `Lidario`. First instantiate a `lidario.Translator` object and set the input type to `geotiff` and the output type to `numpy`. After instantiating the object, use its `lidario.Translator.translate` method with the input set to the raster dataset to generate an array of coordinates. Save the array to your hard drive as a `.npy` NumPy binary with `numpy.save`. With NumPy you can easily open this binary in another Python environment.

```
# import modules
import numpy as np
import lidario as lio

# set path
basename = r'elevation'
dataset = f'{basename}.tif'

# lidario
translator = lio.Translator("geotiff", "numpy")
xyz = translator.translate(dataset)

# export binary
np.save(basename, xyz)
```

To export a raster with a hypsometric color map, save the array with x-, y-, and z-coordinates and another array with red, green, and blue colors as a `.npz` binary. This format can contain multiple arrays such as coordinates and colors. Use the function `numpy.savez_compressed` to save compressed `.npz` binaries. For an example, see the script `geotiff-to-npz.py`.

Code: `geotiff-to-npz.py`

Importing Arrays This example covers how to import a NumPy binary into Rhino and turn it into point cloud geometry using the ScriptEditor.

Data: `elevation.npy`

Code: `npy-to-point-cloud.py`

Start a new Python script in Rhino's ScriptEditor. Import `numpy` and `rhinoscriptsyntax`. Then set the path to the NumPy binary `elevation.npy`. Import the NumPy binary as an array with `numpy.load`. Then generate a point cloud from the array with RhinoScriptSyntax's `AddPointCloud` function. Since the point cloud will be far from the origin and out of view, zoom to the extent of viewports to find it.

```
# requirements: numpy
import numpy as np
import rhinoscriptsyntax as rs

# set path
dataset = 'elevation.npy'

# load array
xyz = np.load(dataset)

# generate point cloud
rs.AddPointCloud(xyz)
```

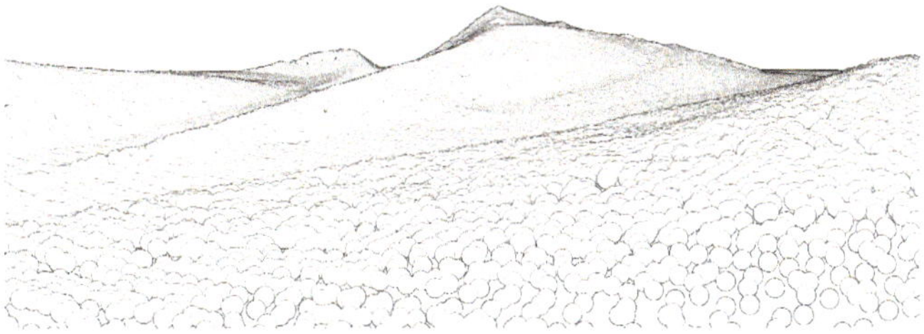

```
Zoom
All
Extents
```

For examples demonstrating how to import a NumPy binary as a point cloud with a hypsometric color map, see the scripts `npy-to-color-cloud.py` and `npz-to-color-cloud.py`.

Code: `npy-to-color-cloud.py` & `npz-to-color-cloud.py`

References

[1] [Software] laspy Development Team, *laspy* version 2.4.1, 2023. lic: laspy license. url: https://laspy.readthedocs.io/en/latest, vcs: https://github.com/laspy/laspy.

[2] [Software] GDAL/OGR Contributors, *GDAL/OGR Geospatial Data Abstraction software Library* version 3.7.1, 2023. Open Source Geospatial Foundation. lic: MIT License. doi: 10.5281/zenodo.8144520, url: https://gdal.org, vcs: https://github.com/OSGeo/gdal.

[3] [Software] GRASS Development Team, *Geographic Resources Analysis Support System (GRASS GIS) Software, Version 8.3* version 8.3.0, 2023. Open Source Geospatial Foundation. lic: GNU General Public License. doi: 10.5281/zenodo.5176030, url: https://grass.osgeo.org, vcs: https://github.com/OSGeo/grass.

[4] Vaclav Petras, Anna Petrasova, James B. McCarter, Helena Mitasova, and Ross K. Meentemeyer. "Point Density Variations in Airborne Lidar Point Clouds". In: *Sensors* 23.3 (Feb. 2023). doi: 10.3390/s23031593.

[5] [Software] Open3D Development Team, *Open3D: A Modern Library for 3D Data Processing* version 0.17, 2023. lic: MIT License. url: http://www.open3d.org/, vcs: https://github.com/isl-org/Open3D.

[6] [Software] PyVista Development Team, *PyVista* version 0.40.0, 2023. lic: MIT License. doi: 10.5281/zenodo.8101943, url: https://pyvista.org/, vcs: https://github.com/pyvista/pyvista.

[7] [Software] Landau Design+Technology, *Land Kit* version 1.2, 2022. url: https://www.landkit.design/, vcs: https://www.food4rhino.com/en/app/land-kit.

[8] [Software] QGIS Development Team, *QGIS Geographic Information System* version 3.32.0, 2022. QGIS Association. lic: GNU General Public License. doi: 10.5281/zenodo.5869838, url: https://www.qgis.org, vcs: https://github.com/qgis/QGIS.

[9] [Software] RhinoTerrain, *RhinoTerrain* version 7.0.272.11, 2022. url: https://www.rhinoterrain.com/.

[10] [Software] Brian Washburn, *Heron* version 0.4.1, 2022. lic: MIT License. vcs: https://github.com/blueherongis/Heron.

[11] [Software] Philip Belesky, *Groundhog* version 0.13.0b, 2021. lic: GNU General Public License v3. url: https://groundhog.la/, vcs: https://github.com/philipbelesky/groundhog.

[12] Katherine R. Barnhart, Eric W.H. Hutton, Gregory E. Tucker, Nicole M. Gasparini, Erkan Istanbulluoglu, Daniel E.J. Hobley, Nathan J. Lyons, Margaux Mouchene, Sai Siddhartha Nudurupati, Jordan M. Adams, et al. "Landlab v2. 0: a software package for Earth surface dynamics". In: *Earth Surface Dynamics* 8.2 (2020), pages 379–397. doi: 10.5194/esurf-8-379-2020.

[13] [Software] Joffrey Bienvenu, *Lidario* version 0.3.0, 2020. lic: MIT License. url: https://lidario.readthedocs.io/, vcs: https://github.com/Joffreybvn/lidario.

[14] Charles R. Harris, K. Jarrod Millman, Stéfan J. van der Walt, Ralf Gommers, Pauli Virtanen, David Cournapeau, Eric Wieser, Julian Taylor, Sebastian Berg, Nathaniel J. Smith, Robert Kern, Matti Picus, Stephan Hoyer, Marten H. van Kerkwijk, Matthew Brett, Allan Haldane, Jaime Fernández del Río, Mark Wiebe, Pearu Peterson, Pierre Gérard-Marchant, Kevin Sheppard, Tyler Reddy, Warren Weckesser, Hameer Abbasi, Christoph Gohlke, and Travis E. Oliphant. "Array programming with NumPy". In: *Nature* 585.7825 (Sept. 2020), pages 357–362. doi: 10.1038/s41586-020-2649-2.

[15] [Software] Eric Hutton, Katherine Barnhart, Dan Hobley, Greg Tucker, Sai Nudurupati, Jordan Adams, Nicole Gasparini, Charlie Shobe, Ronda Strauch, Jenny Knuth, Margaux Mouchene, Nathan Lyons, David Litwin, Rachel Glade, Giuseppecipolla95, Amanda Manaster, Lang-

ston Abby, Kristen Thyng, and Francis Rengers, *landlab* Apr. 2020. lic: MIT. doi: 10.5281/zenodo.595872, url: https://github.com/landlab/landlab.

[16] NASA JPL. *NASADEM Merged DEM Global 1 arc second V001.* distributed by NASA EOSDIS Land Processes DAAC. 2020. doi: 10.5067/MEaSUREs/NASADEM/NASADEM_HGT.001. url: https://doi.org/10.5067/MEaSUREs/NASADEM/NASADEM_HGT.001 (visited on 03/12/2023).

[17] American Society for Photogrammetry and Remote Sensing. *LAS Specification 1.4 – R15.* Technical report. Bethesda, Maryland: ASPRS, July 2019. url: https://www.asprs.org/wp-content/uploads/2019/07/LAS_1_4_r15.pdf (visited on 01/03/2023).

[18] Open Geospatial Consortium. *OGC GeoTIFF Standard.* 2019. url: https://www.ogc.org/standard/geotiff/ (visited on 03/10/2023).

[19] European Space Agency. *Copernicus DEM – Global and European Digital Elevation Model (COP-DEM).* 2019. doi: https://doi.org/10.5270/ESA-c5d3d65. url: https://spacedata.copernicus.eu/collections/copernicus-digital-elevation-model/ (visited on 03/12/2023).

[20] Ilmar Hurkxkens and Mathias Bernhard. "Computational Terrain Modeling with Distance Functions for Large Scale Landscape Design". In: *Journal of Digital Landscape Architecture* 2.4 (2019), pages 222–230. doi: 10.14627/537663024.

[21] [Software] Ilmar Hurkxkens and Mathias Bernhard, *Docofossor* version 0.904, 2019. lic: MIT License. vcs: https://github.com/dbt-ethz/docofossor.

[22] Bane Sullivan and Alexander Kaszynski. "PyVista: 3D plotting and mesh analysis through a streamlined interface for the Visualization Toolkit (VTK)". In: *Journal of Open Source Software* 4.37 (May 2019), page 1450. doi: 10.21105/joss.01450.

[23] [Software] Bison Development Team, *Bison* version 0.9, 2018. url: https://www.food4rhino.com/en/app/bison.

[24] Dai Yamazaki. *MERIT DEM: Multi-Error-Removed Improved-Terrain DEM.* 2018. url: http://hydro.iis.u-tokyo.ac.jp/~yamadai/MERIT_DEM/ (visited on 03/12/2023).

[25] Qian-Yi Zhou, Jaesik Park, and Vladlen Koltun. "Open3D: A Modern Library for 3D Data Processing". In: *arXiv:1801.09847* (2018). doi: 10.48550/arXiv.1801.09847.

[26] Igor V. Florinsky. "An illustrated introduction to general geomorphometry". In: *Progress in Physical Geography* 41.6 (2017), pages 723–752. issn: 03091333. doi: 10.1177/0309133317733667.

[27] Noel Gorelick, Matt Hancher, Mike Dixon, Simon Ilyushchenko, David Thau, and Rebecca Moore. "Google Earth Engine: Planetary-scale geospatial analysis for everyone". In: *Remote Sensing of Environment* (2017). doi: 10.1016/j.rse.2017.06.031.

[28] Daniel E.J. Hobley, Jordan M. Adams, Sai Siddhartha Nudurupati, Eric W.H. Hutton, Nicole M. Gasparini, Erkan Istanbulluoglu, and Gregory E. Tucker. " Creative computing with Landlab: an open-source toolkit for building, coupling, and exploring two-dimensional numerical models of Earth-surface dynamics ". In: *Earth Surface Dynamics* 5.1 (2017), pages 21–46. doi: 10.5194/esurf-5-21-2017.

[29] New York State GIS Resources. *NYS DEM: NYC Topobathymetric Data.* 2017. url: https://gis.ny.gov/nys-dem (visited on 08/16/2023).

[30] Dai Yamazaki, Daiki Ikeshima, Ryunosuke Tawatari, Tomohiro Yamaguchi, Fiachra O' Loughlin, Jeffery C. Neal, Christopher C. Sampson, Shinjiro Kanae, and Paul D. Bates. "A high-accuracy map of global terrain elevations". In: *Geophysical Research Letters* 44.11 (2017), 5844–5853. doi: 10.1002/2017GL072874.

[31] U.S. Geological Survey. *3D Elevation Program.* 2016. url: https://www.usgs.gov/3d-elevation-program (visited on 03/12/2023).

[32] West 8. *Governor's Island: The Hills*. 2016. url: https://www.west8.com/projects/governors_island_phase_2_the_hills (visited on 03/05/2023).

[33] Japan Aerospace Exploration Agency. *ALOS Global Digital Surface Model*. 2014. url: https://www.eorc.jaxa.jp/ALOS/en/dataset/aw3d30/aw3d30_e.htm (visited on 03/05/2023).

[34] Takeo Tadono, Haruyuki Ishida, Fumiko Oda, Soichiro Naito, Keiji Minakawa, and Hisanori Iwamoto. "Precise Global DEM Generation by ALOS PRISM". In: *ISPRS Annals of Photogrammetry, Remote Sensing and Spatial Information Sciences* II4 (Apr. 2014), pages 71–76. doi: 10.5194/isprsannals-II-4-71-2014.

[35] [Software] Sean Gillies et al., *Rasterio: geospatial raster I/O for Python programmers* 2013. Mapbox. url: https://github.com/rasterio/rasterio.

[36] Helena Mitasova, Michael Barton, Isaac Ullah, Jaro Hofierka, and Russell S. Harmon. "3.9 GIS-Based Soil Erosion Modeling". In: *Treatise on Geomorphology*. Edited by John F. Shroder. San Diego: Academic Press, 2013, pages 228–258. isbn: 978-0-08-088522-3. doi: 10.1016/B978-0-12-374739-6.00052-X.

[37] Thad Wasklewicz, Dennis Staley, Kathryn Reavis, and Takashi Oguchi. "3.6 Digital Terrain Modeling". In: *Treatise on Geomorphology*. Edited by John F. Shroder. San Diego: Academic Press, 2013, pages 130–161. isbn: 978-0-08-088522-3. doi: 10.1016/B978-0-12-374739-6.00048-8.

[38] OpenTopography Facility. *OpenTopography*. 2008. url: https://opentopography.org/ (visited on 03/12/2023).

[39] Jeffrey S. Evans and Andrew T. Hudak. "A Multiscale Curvature Algorithm for Classifying Discrete Return LiDAR in Forested Environments". In: *IEEE Transactions on Geoscience and Remote Sensing* 45.4 (2007), pages 1029–1038. doi: 10.1109/TGRS.2006.890412.

[40] Tom G. Farr, Paul A. Rosen, Edward Caro, Robert Crippen, Riley Duren, Scott Hensley, Michael Kobrick, Mimi Paller, Ernesto Rodriguez, Ladislav Roth, David Seal, Scott Shaffer, Joanne Shimada, Jeffrey Umland, Marian Werner, Michael Oskin, Douglas Burbank, and Douglas Alsdorf. "The Shuttle Radar Topography Mission". In: *Reviews of Geophysics* 45.2 (2007). doi: https://doi.org/10.1029/2005RG000183.

[41] Helena Mitasova, Lubos Mitas, and Russell S. Harmon. "Simultaneous spline approximation and topographic analysis for lidar elevation data in open-source GIS". In: *IEEE Geoscience and Remote Sensing Letters* 2.4 (2005), pages 375–379. doi: 10.1109/LGRS.2005.848533.

[42] Maya Lin. *Eleven Minute Line*. 2004. url: https://www.mayalinstudio.com/art/eleven-minute-line (visited on 03/05/2023).

[43] NASA Jet Propulsion Laboratory. *Shuttle Radar Topography Mission*. 2003. url: https://www2.jpl.nasa.gov/srtm/ (visited on 03/12/2023).

[44] Google. *Google Earth Engine*. 2002. url: https://earthengine.google.com/ (visited on 03/12/2023).

[45] Charles Jencks. *Landform Ueda*. 2002. url: https://www.charlesjencks.com/projects-ueda (visited on 03/05/2023).

[46] Lubos Mitas and Helena Mitasova. "Spatial Interpolation". In: *Geographical Information Systems: Principles, Techniques, Management and Applications*. Edited by Paul A. Longley, Michael F. Goodchild, David J. Maguire, and David W. Rhind. Wiley, 1999, pages 481–492. isbn: 978-0-471-73545-8.

[47] Der-Tsai Lee and Bruce J. Schachter. "Two algorithms for constructing a Delaunay triangulation". In: *International Journal of Computer & Information Sciences* 9.3 (1980), pages 219–242.

[48] Boris Delaunay. "Sur la sphère vide". In: *Bulletin de l'Académie des Sciences de l'URSS. Classe des sciences mathématiques et na* 1934.6 (1934), pages 793–800.

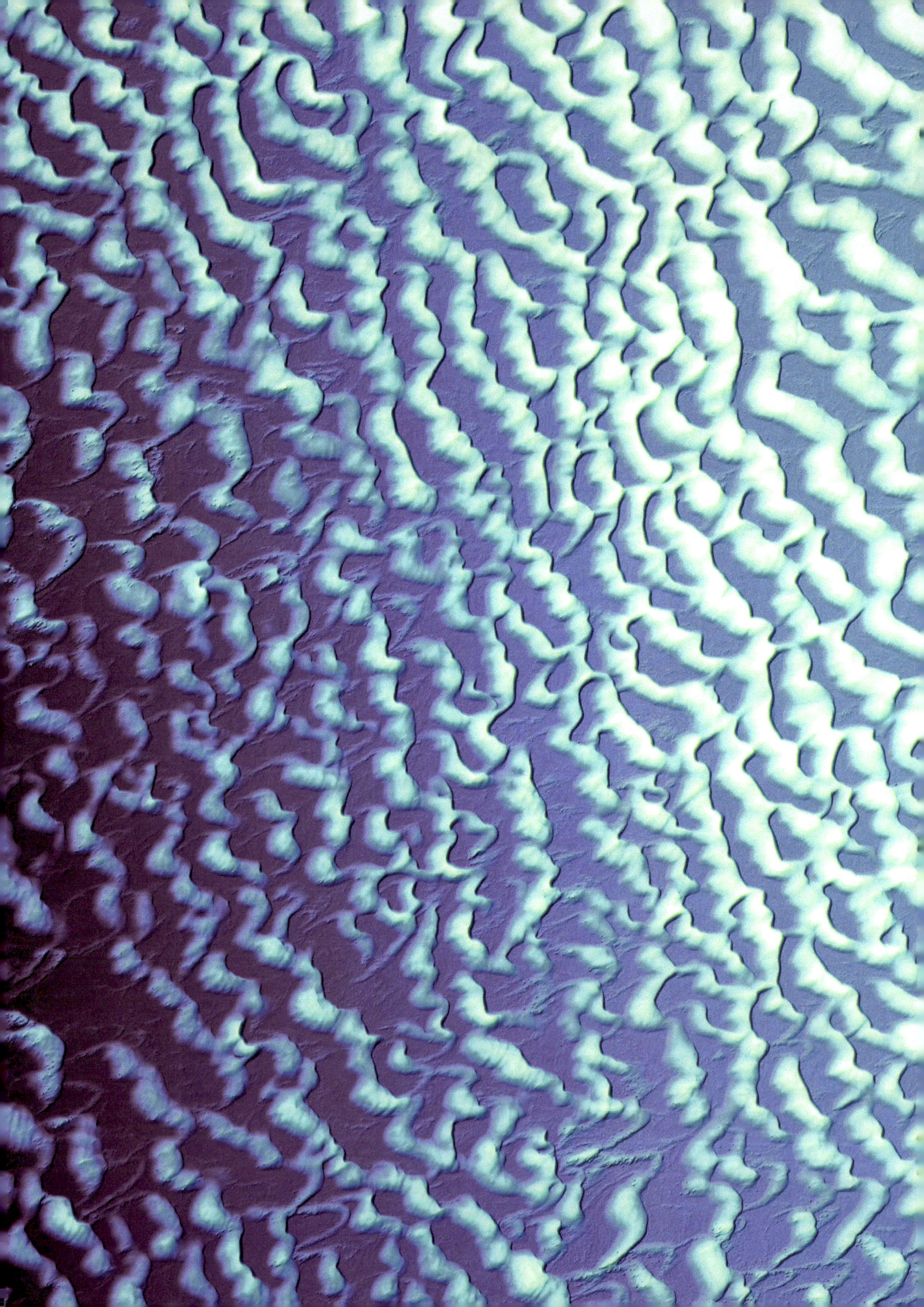

16 Dunes

Import and model the dunes at White Sands National Monument in New Mexico. The White Sands dune field is composed primarily of crescent-shaped barchan and parabolic dunes. The dunes are shaped by Aeolian processes – the brilliant white gypsum sand is eroded, transported, and deposited by predominately northeasterly winds. This arid, windswept landscape is dynamic. The crest lines of the barchan dunes evolve – decomposing and reforming – on a decadal timescale, while the parabolic dunes on the periphery can migrate up to 2.4 meters every year [2].

16.1 Lidar

Download elevation data for the White Sands dune field from Open Topography. A 2010 airborne lidar survey of the dune fields is hosted at https://doi.org/10.5069/G97D2S2D. Browse to `Point Cloud Data`. To define your study region, check `manually enter selection coordinates` and then set `Xmin`, `Ymin`, `Xmax`, and `Ymax` as shown below. This will set a 500 by 500 meter region containing approximately 1.3 million points with a point density of 4.62 points per square meter. The lidar data can be downloaded from Open Topography as an ASCII point cloud, a LAZ point cloud, or a raster. These datasets can also be downloaded from the data repository for this book at https://doi.org/10.5281/zenodo.8212063.

```
Xmin = 383388
Ymin = 3634200
Xmax = 383888
Ymax = 3634700
```

ASCII First download a plain text ASCII point cloud for modeling in Grasshopper. Under the heading `Point Cloud Data Download`, select `ASCII` format and submit your job. When the job is ready, download and unarchive `points.tar.gz`. This archive contains the ASCII point cloud `points.txt`. Use a free, open source utility such as 7-Zip for unarchiving.

LAZ Download a `.laz` point cloud for modeling in Grasshopper with RhinoCode. Under the heading `Point Cloud Data Download`, select `LAZ` format and submit your job.

Raster Download a `.tif` raster for plotting with Python. Open Topography can interpolate a digital elevation model from the point cloud by triangulation or binning. This example uses binning, i.e. local gridding. Under the heading `DEM Generation (Local Gridding)`, select `Calculate Zmean grid`, set the grid resolution to 1 meter, and submit your job.

16.2 Grasshopper

Import the ASCII point cloud of the White Sands dune field into Rhino. Then triangulate a mesh between the points with Grasshopper.

Data: `points.txt`

Code: `white-sands.gh`

Import Import the ASCII point cloud `points.txt` into Rhino. In the command line, run the command `Import`, set the delimiter to comma, and check `create point cloud`. The imported point cloud will be far from the origin of your scene and thus outside of your current view, so zoom to the point cloud with the `Zoom` command. The point cloud should have about 1.3 million points. Grasshopper will struggle to triangulate a mesh through this many points. Furthermore, the point cloud has artifacts from the scanning process that would make the resulting mesh appear noisy. If you zoom in on the cloud point in top view, a zig-zag pattern will appear. When airborne lidar is captured with a swinging or oscillating mirror, the point cloud will have a highly variable point density and a distinctive zig-zag pattern [1]. Decimate the point cloud with the `ReducePointCloud` command to make it a manageable size and reduce variations in point density. Run the command `ReducePointCloud` and set the number of points to keep to 250,000. To better visualize the terrain, consider scaling the point cloud along the z-axis. Vertical exaggeration will accentuate landforms. To vertically exaggerate the terrain, select the point cloud and run the command `Scale1D`. Set the scale factor to 2 and the scale direction to the z-axis. Since Grasshopper's `Delaunay Mesh` component takes points rather than a point cloud as its input, explode the point cloud with the `Explode` command.

```
Import
Zoom
ReducePointCloud
Scale1D
Explode
```

Delaunay Triangulation Triangulate a mesh from the points in Grasshopper. In Grasshopper add a `Point` parameter to the canvas. Right click on the parameter to set multiple points. In Rhino select the points from the point cloud. Triangulate a mesh through the points with the `Delaunay Mesh` component. Set a color map for the elevation gradient of the mesh. First deconstruct the mesh with the `Deconstruct Mesh` component. Then find the elevation gradient of the mesh by deconstructing its vertices and extracting their z-value with the `Deconstruct` component. Remap the z-values to a domain from 0 to 1 for use as a color gradient. To do this, plug the z-values into a `Remap Numbers` component and set its source domain to the `Bounds` of the z-values. Plug the remapped z-values into a `Gradient` component. Finally reconstruct the mesh with the `Construct Mesh` component using the new color gradient. Experiment with color gradients from `color-maps.gh`.

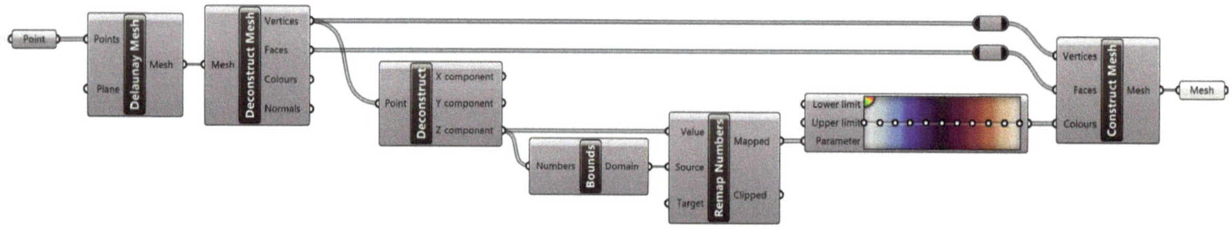

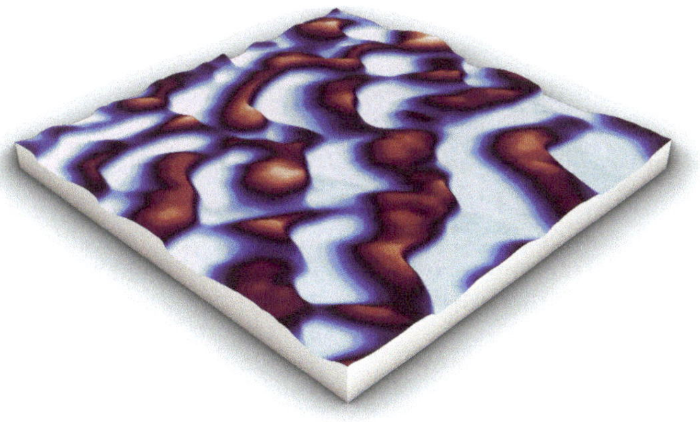

16.3 Python

Using Python, plot the raster of the White Sands dune field. First import `rasterio` and `rasterio.plot.show`. Then open the digital elevation model with `rasterio.open` and plot it with `rasterio.plot.show`. Experiment with perceptually uniform color palettes such as mako, twilight, and icefire.

Data: `output.mean.tif`
Code: `white-sands-raster.py`

```
# import modules
import rasterio as rio
from rasterio.plot import show

# set path
dataset = r'output.mean.tif'

# read raster
elevation = rio.open(dataset)

# plot raster
show(elevation, cmap='mako')
```

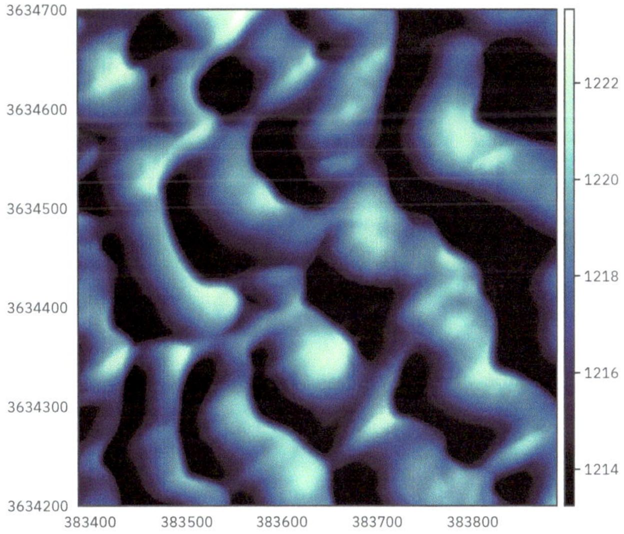

16.4 Rhino Scripting

Use Rhino's Script Editor to import, decimate, and transform a `.laz` lidar point cloud of the dune field. Write a Python script that uses the Laspy library to import a LAZ point cloud. Then use slicing to decimate the point cloud, reducing its size, and removing noise. Use RhinoScriptSyntax to generate, transform, and visualize the point cloud as geometry in Rhino.

Data: `points.laz`

Code: `white-sands-lidar.py`

Libraries In the Script Editor create a new Python script. Set `lazrs` and `laspy` as required packages. Then import `laspy` and `rhinoscriptsyntax`.

Read Set the file path to the dataset `points.laz` and then read it with Laspy's `laspy.read` method. This will return a `laspy.LasData` object. The point cloud's coordinates can be accessed by calling the object's `.xyz` attribute. This will return an array of x-, y-, and z-values.

Decimate Decimate the point cloud by slicing this array of coordinates. Python's slicing notation is `[start:stop:step]`. The statement `xyz = las.xyz[::10]` will return every tenth set of coordinates, removing 90 percent of the points.

Generate Use RhinoScriptSyntax's `AddPointCloud` function to create a point cloud in Rhino from the array of coordinates.

Scale For the sake of visualization, try vertically exaggerating the point cloud to accentuate the shape of the landforms. Use RhinoScriptSyntax's `ScaleObject` function to scale the point cloud on the z-axis. Use slicing to set the origin of scaling to the first point in the point cloud. Set the scale factor to 1 on x-axis, 1 on the y-axis, and 2 on the z-axis.

```
# requirements: lazrs, laspy
import laspy
import rhinoscriptsyntax a

# set path
data = r'points.laz'

# read lidar
las = laspy.read(data)

# decimate point cloud
xyz = las.xyz[::10]

# generate point cloud
cloud = rs.AddPointCloud(x

# scale point cloud
origin = xyz[0]
scale = (1, 1, 2)
rs.ScaleObject(cloud, origin, scale)
```

References

[1] Vaclav Petras, Anna Petrasova, James B. McCarter, Helena Mitasova, and Ross K. Meentemeyer. "Point Density Variations in Airborne Lidar Point Clouds". In: *Sensors* 23.3 (Feb. 2023). doi: 10.3390/s23031593.

[2] Katie KellerLynn. *White Sands National Monument: Geologic Resources Inventory Report*. Natural Resource Report NPS/NRSS/GRD/NRR—2012/585. Fort Collins, Colorado: National Park Service, Oct. 2012. url: https://irma.nps.gov/DataStore/Reference/Profile/2190397.

17 Earthworks

17.1 Grading

Grading is the design of earthworks, of topographic gradients. Landscape architects reshape terrain, grading paths and roads with gentle slopes, leveling ground for buildings, and building new landforms such as mounds and ridges [17]. Terrain can be designed as expressive, tectonic forms like Hargreaves Jones' South Pointe Park [19] and West 8's Hills of Governors Island [14]. These landforms are geomorphic metaphors, abstract artistic references to physical processes. Terrain can be designed for performance like Snøhetta's MAX IV Laboratory Landscape, which dampens highway vibrations [13], or H+N+S' Park Buitenschot, which dampens noise from Schiphol Airport [15]. Terrain can also be designed as a catalyst for processes like the Sand Motor [18, 16] or the River Aire [12]. Programs for designing earthworks include Civil3D [1] and Landmark [2]. Grasshopper also has plugins for designing earthworks including Bison [11], Landkit [3], and Docofossor [9]. Earthmoving operations can be classified as cut or fill, as excavation or embankment. For cutting operations, earth is removed – either dug out or pushed away. For fill operations, earth is added – either deposited or pushed in place. In this chapter we will model cut-and-fill operations with Docofossor. Download and install this plugin from https://www.food4rhino.com/en/app/docofossor.

Plugin: `Docofossor`

17.1.1 Volumetric Modeling

Terrain is usually represented computationally as explicit geometry such as points or meshes that are defined by coordinates in space. Terrain, however, can also be represented implicitly by mathematical functions. With volumetric modeling based on signed distance functions, terrain can be represented as a smooth, continuous, mathematically defined surface that can easily be transformed by Boolean operations [8]. With signed distance functions, points are no longer defined explicitly by a set of coordinates, but rather implicitly by a distance from a shape [10]. With volumetric modeling, terrain can be graded intuitively and efficiently using Boolean addition and subtraction for cut-and-fill operations.

`Docofossor` is an addon for volumetric terrain modeling in Grasshopper [9, 8, 6]. Docofossor uses signed distance functions to represent terrain as a continuous field of elevation values that can be modeled with Boolean operations. As a distance field `df`, terrain is represented as height above a datum such as mean sea level. In Docofossor a distance field is a grid of elevation values above a datum (Figure 17.1). The data structure `df` is a list with a header describing the grid followed by a sequence of z-values. The z-values are listed column by column from the bottom left. The distance field can be transformed by multiple Boolean operations to regrade the terrain. Boolean addition will push points in the grid up, while Boolean subtraction will push points down. In terms of grading, Boolean addition is filling, while Boolean subtraction is cutting, i.e. excavating. These Boolean operations are not computationally expensive because they are solved arithmetically as the difference between distance fields. This approach – volumetric modeling with Docofossor – has been used to model complex, computationally generated terrain [7, 4] and design an autonomously constructed embankment [5].

Columns	101	
Rows	101	
Local Origin X	0	
Local Origin Y	0	
Cell Size X	1	Header
Cell Size Y	1	
	0	
	0	
Global Origin X	0	
Global Origin Y	0	
	1.1433	
	1.7488	
	2.351	Elevation
	2.9495	
	...	

Figure 17.1: Distance field

17.1.2 Operations

In volumetric modeling cut-and-fill operations are calculated with either Boolean subtraction or addition. These operations can be either relative or absolute. In absolute operations earth is either added or removed until a specific elevation is reached. In relative operations, however, earth is added or removed by a depth relative to the existing terrain. With Docofossor the depth of cut or fill is designated with geometry – with either points, curves, or surfaces. Points can be used to mark the lowest or highest elevation of peaks or depressions. Curves can be used to draw ridgelines or thalwegs for ridges or valleys. They can also be used to outline boundaries for cut or fill. Key parameters include the position of the geometry `pt` or `crv`, the width of the path `w`, the depth of the operation `mxd`, and the angle of the slope `sa`. In this chapter, we explore five key volumetric modeling operations: cutting in points, filling in points, cutting and filling in paths, cutting on paths, and filling on paths. Complex landforms can be modeled by combining these basic operations.

Code: `cut-fill-operations.gh`

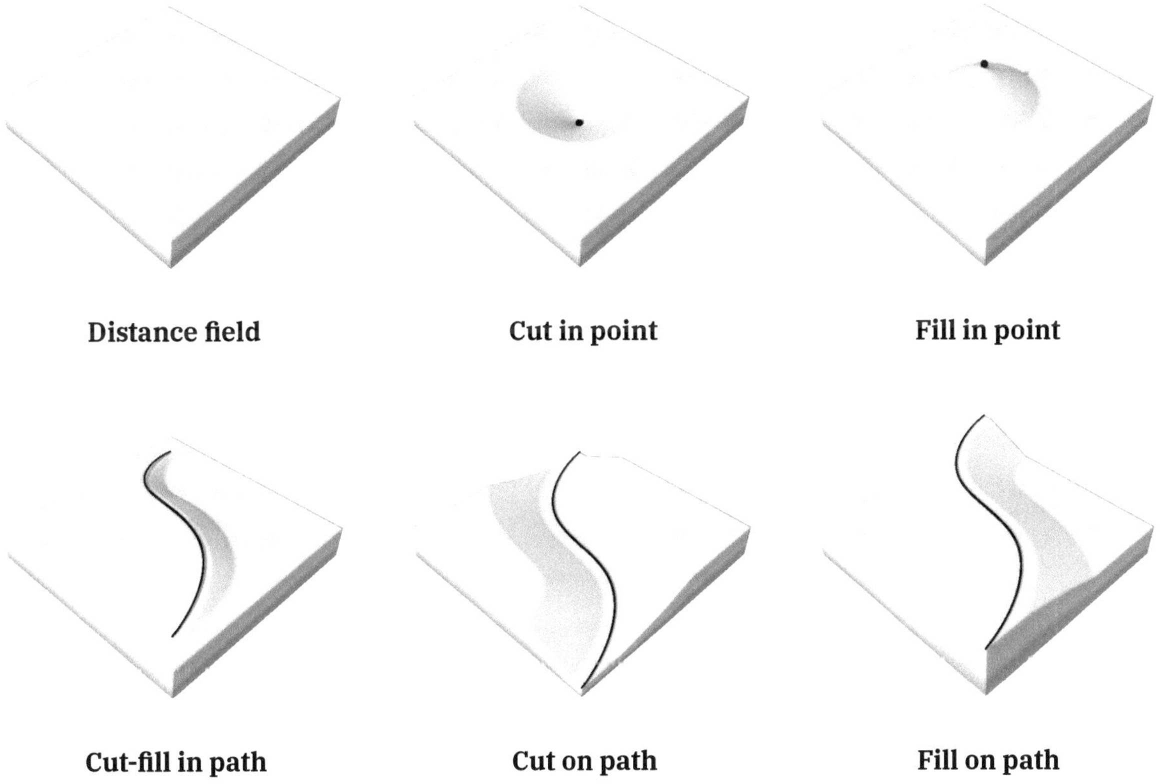

Figure 17.2: Cut-fill operations

Distance Field Distance fields can be generated by importing .xyz point clouds, .asc rasters, or Grassshopper points. Distance fields can also be generated as empty grids with the dfEmptyGrid component. In this example, we will generate a distance field from a square grid of Grasshopper points, rather than an empty grid. Add a Square component and set the cell size and extent of the grid using number sliders. Flatten the output tree of points into a list. Plug the list of points into a dfImportPoints component to create a distance field df. Visualize the resulting distance field as a mesh with a dfGridMesh component.

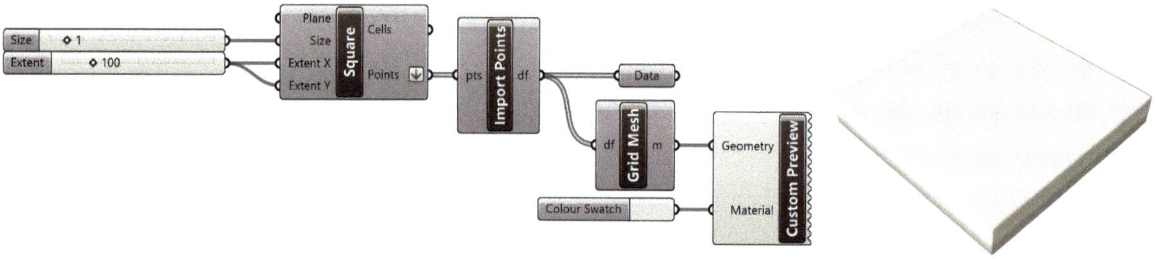

Cut in Point Use the dfCutInPoint component in Docofossor's absolute operations tab to excavate a pit at a specific point. First add a Point parameter and set it to coordinates beneath the terrain. Optionally right click on this point parameter and internalize the point. This will store the point in the Grasshopper definition and enable transformations with the gumball. Try using the gumball to move the point.

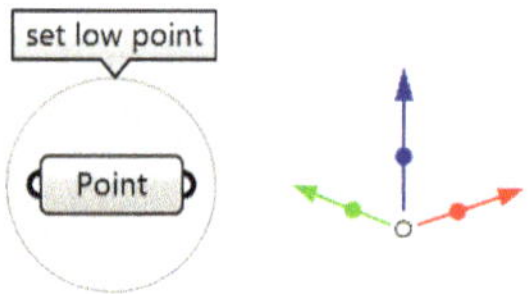

Add a dfCutInPoint component to the canvas. Set the input distance field df to the distance field generated from a square grid of points. Set the input point pt to the low point. Set the slope angle sa and the maximum cutting distance d with number sliders. Visualize the result as a mesh with a dfGridMesh component.

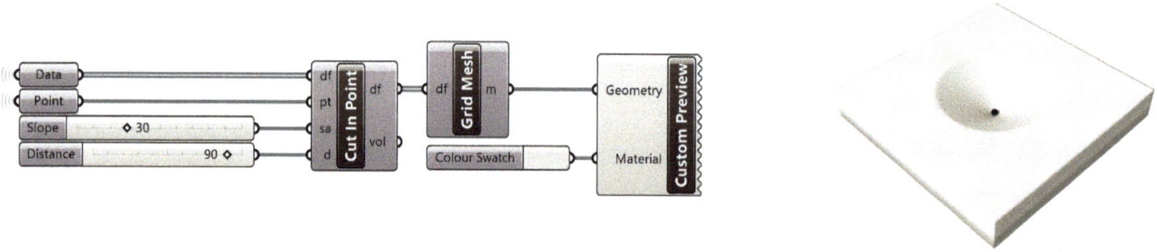

Fill in Point Use the dfFillInPoint component in Docofossor's absolute operations tab to build a mound at a specific point. First add a Point parameter and set it to coordinates above the terrain.

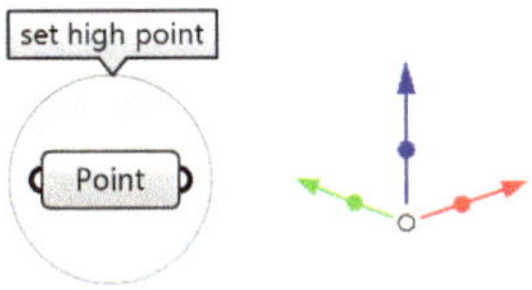

Add a dfFillInPoint component to the canvas and set the input distance field df. Then set the input point pt to the high point. Set the slope angle sa and maximum filling distance d with number sliders. Visualize the result as a mesh with a dfGridMesh component.

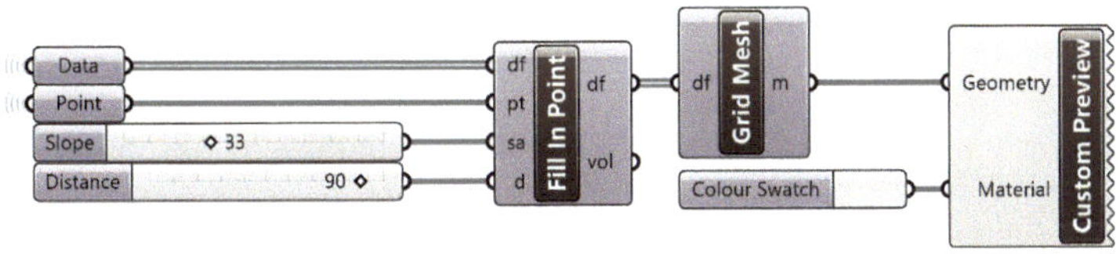

Cut-Fill in Path Use the dfCutFillInPath component in Docofossor's absolute operations tab to construct a landform along a curve. Start by drawing a 3D curve that will be the ridgeline of the landform. Store this in a Curve parameter.

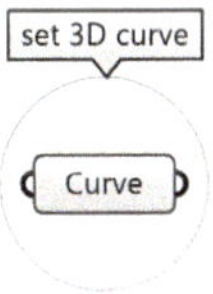

Add a dfCutFillInPath component to the canvas and set the input distance field df. Then set the input curve crv to the 3D curve that models the ridgeline. Set the width of the path w, the slope angle sa, and the maximum distance d with number sliders. Visualize the result as a mesh with a dfGridMesh component.

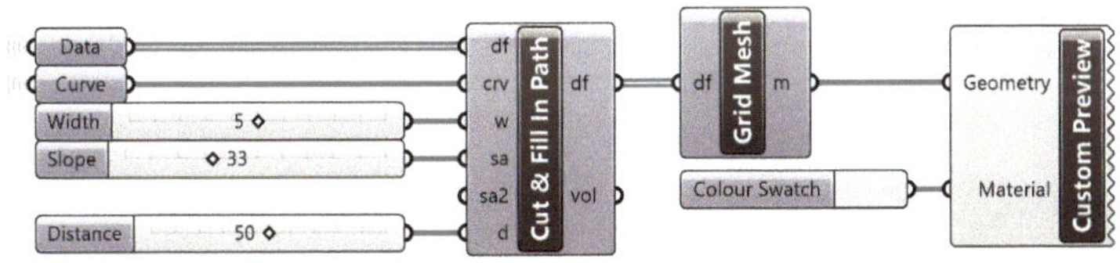

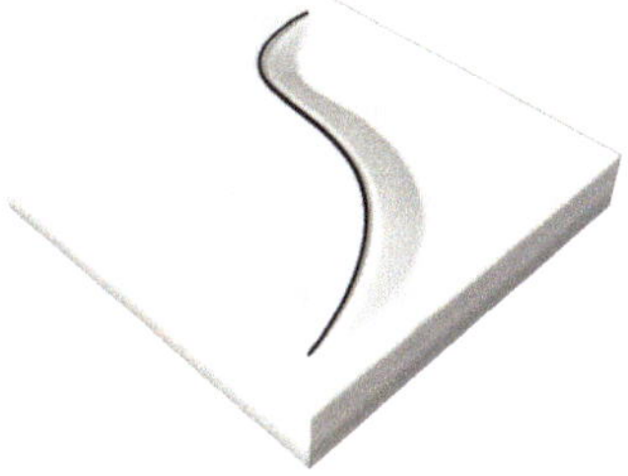

Cut on Path Use the `dfCutOnPath` component in Docofossor's relative operations tab to excavate by a given depth along a curve. Start by drawing a 2D curve and storing it in a `Curve` parameter. The height of the curve does not matter as the cutting depth is relative to the terrain.

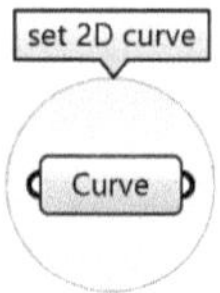

Add a `dfCutOnPath` component to the canvas and set the input distance field `df`. Then set the input curve `crv` to the 2D curve. Set the maximum cutting depth `mxd`, the width of the path `wb`, the slope angle `sa`, and maximum cutting distance `d` with number sliders. Visualize the result as a mesh with a `dfGridMesh` component.

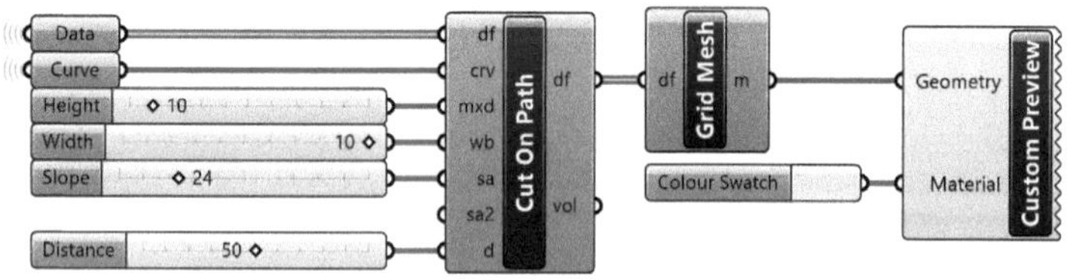

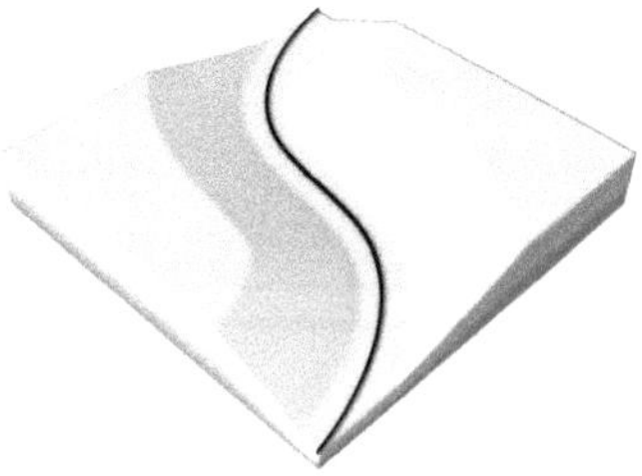

Fill on Path Use the `dfFillOnPath` component in Docofossor's relative operations tab to build an embankment of a given height along a curve. Start by drawing a 2D curve and storing it in a `Curve` parameter. Again the height of the curve does not matter as the embankment height is relative to the terrain.

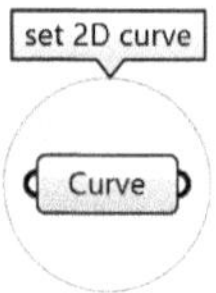

Add a `dfFillOnPath` component to the canvas and set the input distance field `df`. Then set the input curve `crv` to the 2D curve. Set the maximum filling height `mxh`, the width of the path `wt`, the slope angle `sa`, and maximum filling distance `d` with number sliders. Visualize the result as a mesh with a `dfGridMesh` component.

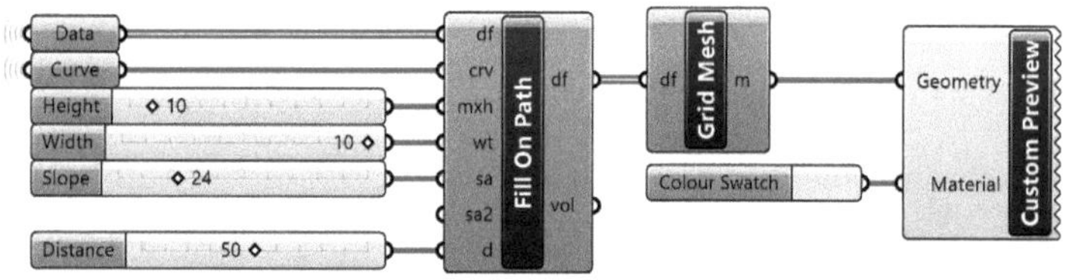

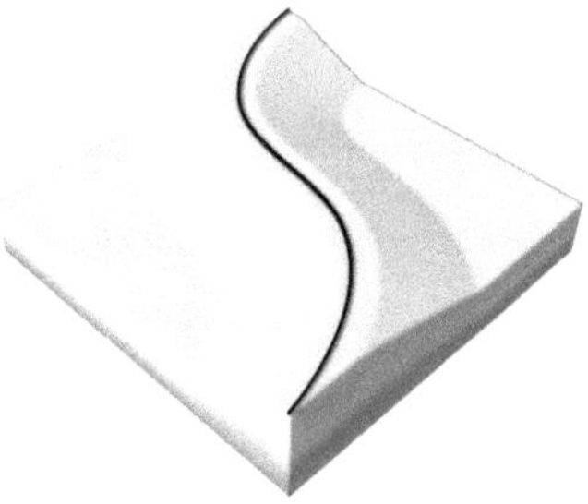

17.2 Modeling Landforms

Use volumetric modeling to design a new landform for the Hills of Governor's Island. Import the dataset `governors-island-landform.xyz`, model a new landform by cutting and filling in a path, calculate the volume of cut-and-fill, derive contours, visualize the scene, and export the edited dataset.

Docofossor can import gridded `.xyz` point clouds. The simplest way to prepare a gridded dataset for Docofossor is to export an elevation raster as an ASCII text file. If you have an unstructured point cloud rather than a raster, use CloudCompare's `Rasterize` tool to bin the point cloud into a grid. This can also be done in Python with `CloudComPy`'s method `cloudComPy.RasterizeToCloud`. Docofossor expects a regular grid in column-major order starting from the lower left corner. This is a major limitation as many topographic datasets are stored differently. A simple solution is simply to mirror the final mesh modeled with Docofossor along the xz-plane.

Code: `grading.gh`

Data: `governors-island-landform.xyz`

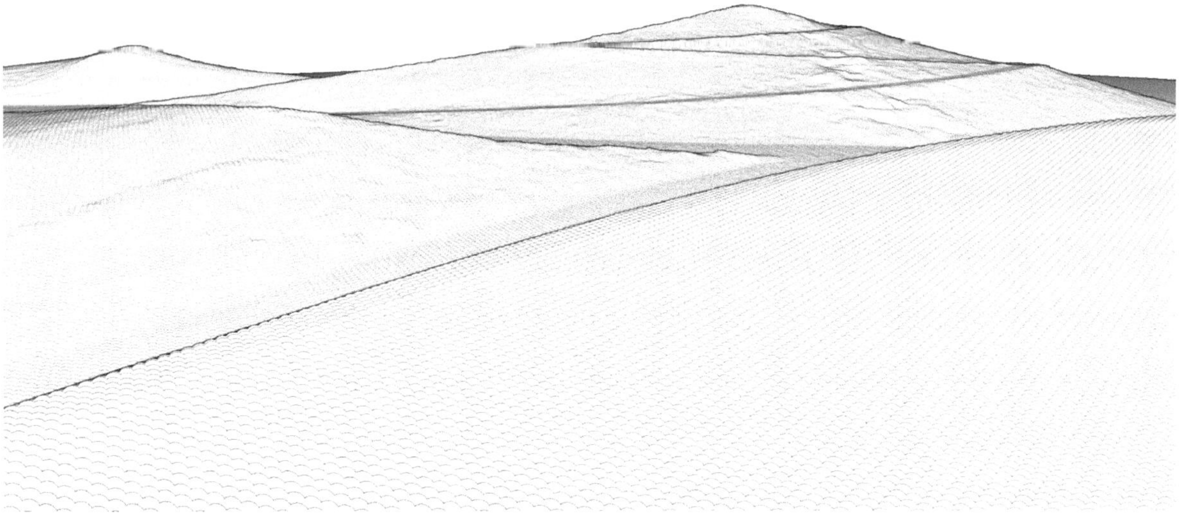

Figure 17.3: Hills of Governor's Island

Modeling To generate a terrain model of the Hills of Governor's Island, import a gridded ASCII point cloud as a distance field and visualize it as a mesh. Start by adding a file path parameter. Right click on it and set the path to the dataset `governors-island-landform.xyz`. Plug this into a `dfImportXYZ` component as the input file path `f`. Set the resolution `n` with a number slider. This will load the `.xyz` point cloud as a signed distance field. For ease of modeling, shift the distance field from its global coordinate system `NAD83 / New York Long Island (ftUS)` to a local coordinate system. The origin of the grid will be shifted to the origin of the scene at `0,0,0`. Visualize the distance field as a mesh with a `dfGridMesh` component.

Global origin	**Local origin**
`976850.5,189100.5,0`	`0,0,0`

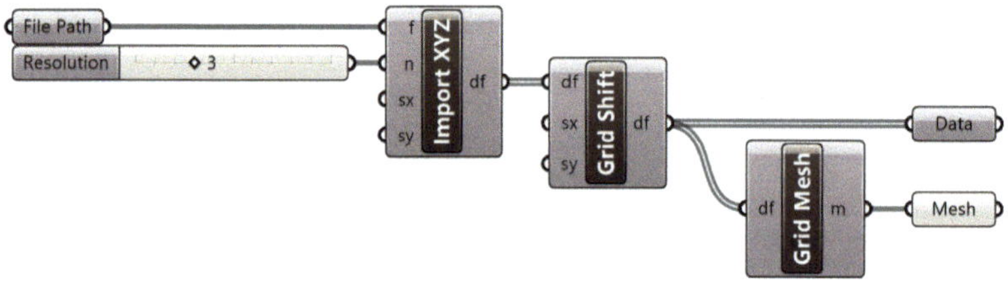

Grading Use volumetric modeling to design a new landform for the Hills of Governor's Island. Start by drawing a 3D curve representing the ridgeline of the new landform. To draw an easily editable curve, make a set of points in Rhino, store and internalize them in a point parameter, and then interpolate a curve through these points with an `Interpolate` component. To reshape the interpolated curve, simply move the internalized points using their gumballs. A gumball is a tool for intuitively transforming geometry. When you select a point, you can use its gumball to move it along the x-, y-, or z-axis. Activate gumballs in Grasshopper in the `Display` menu. Once geometry has been internalized, its gumball will appear.

Use an absolute cut-and-fill operation to construct a ridge along the interpolated curve. Add a `dfCutFillInPath` component to the canvas and set the input distance field `df`. Then set the input curve `crv` to the interpolated curve. Set the width of the path `w`, the outer slope angle `sa`, the inner slope angle `sa2`, and the maximum distance `d` with number sliders. Visualize the result as a mesh with a `dfGridMesh` component.

Gumball

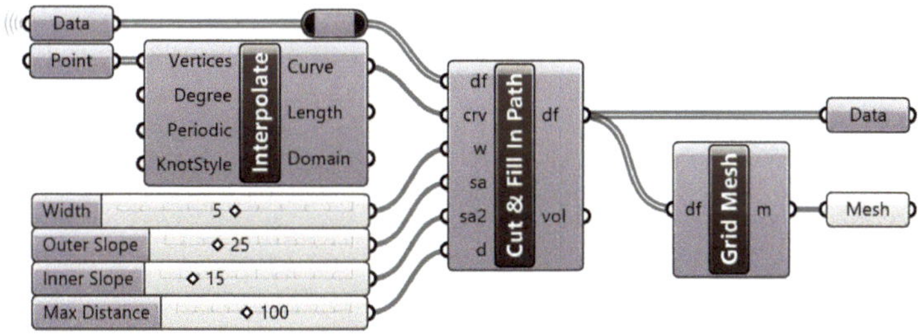

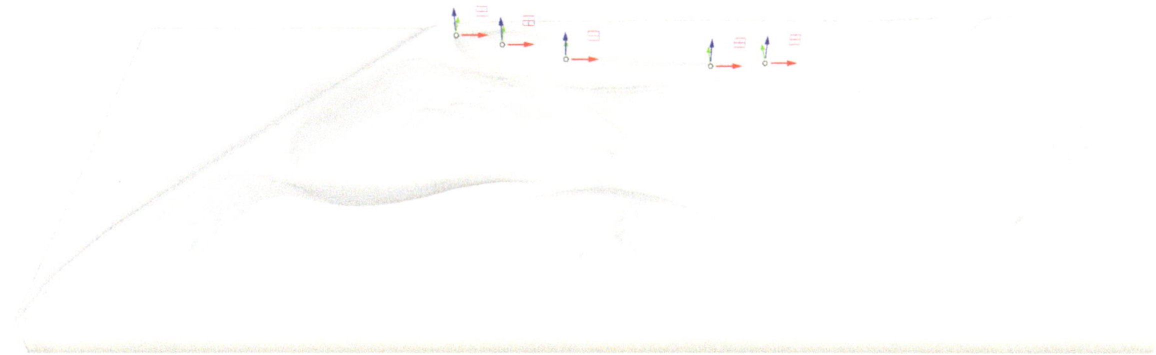

Calculation Calculate the volume of cut-and-fill with a `dfGridCompare` component. Set the first input distance field `df1` to the original terrain. Set the second input distance field `df2` to the edited terrain with the new landform. Display the output volumes of cut-and-fill on text panels. Visualize the volume of cut-and-fill by plugging the output distance field `df` into a `dfGridMesh` component.

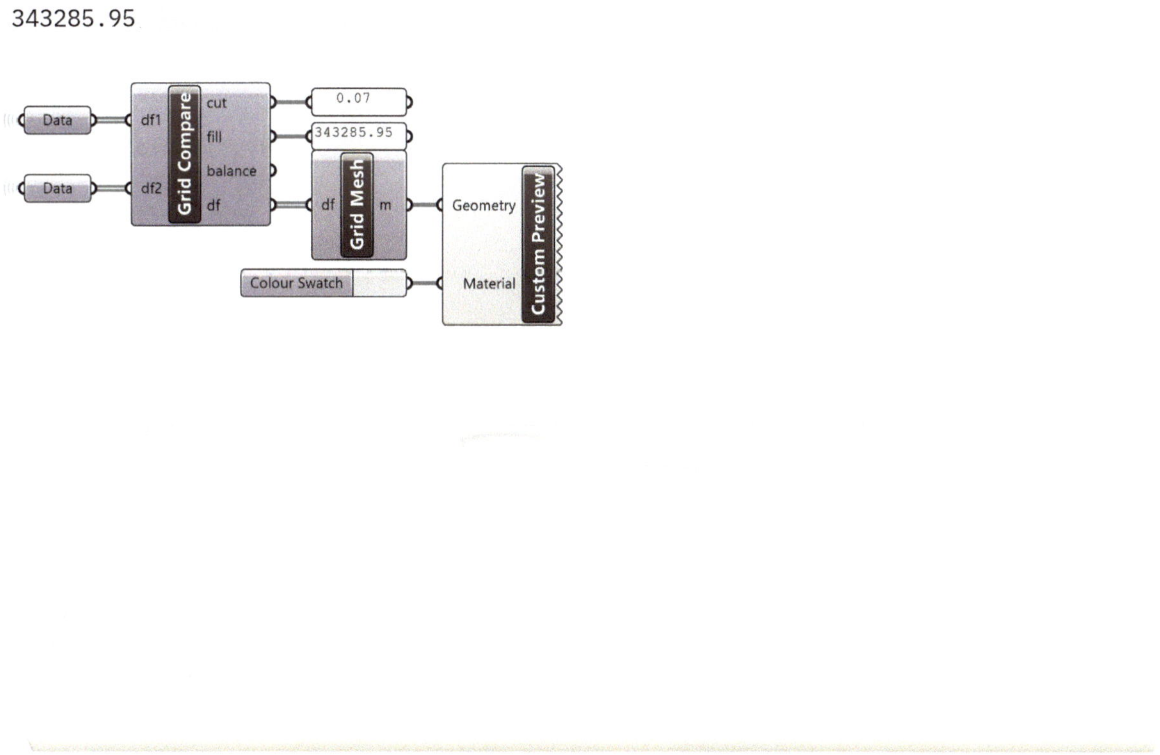

Visualization Since Docofossor expects input data in column-major order, but the Hills of Governor's Island dataset is in row-major order, the mesh needs to be mirrored along the xz-plane. To mirror the terrain, plug the mesh into a `Mirror` component. Set the mirror plane to the `XZ Plane`. Calculate the origin of the xz-plane, by finding the centroid of the terrain mesh using an `Area` component. Use a `Custom Preview` with a `Colour Swatch` to visualize the mirrored mesh.

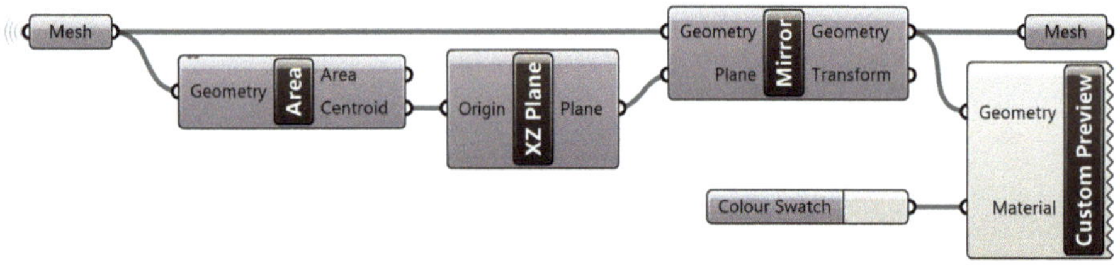

Contours Compute contours from the edited terrain mesh. Add a `Contour` component and set the input shape to the edited terrain mesh, set the direction to the `Unit Z` vector, and set the distance between contours with a number slider. Visualize the contours using a `Custom Preview` with a `Colour Swatch`.

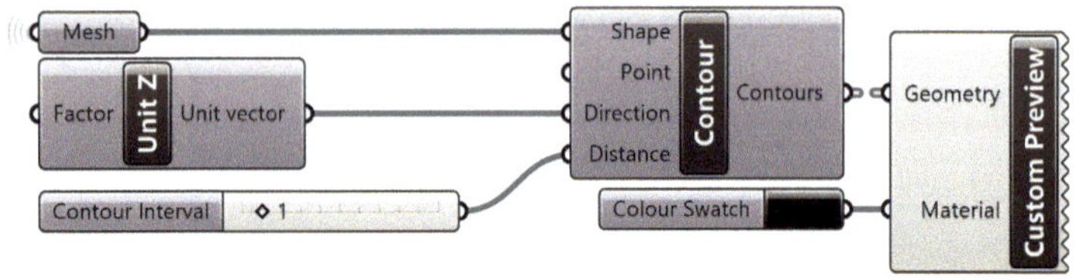

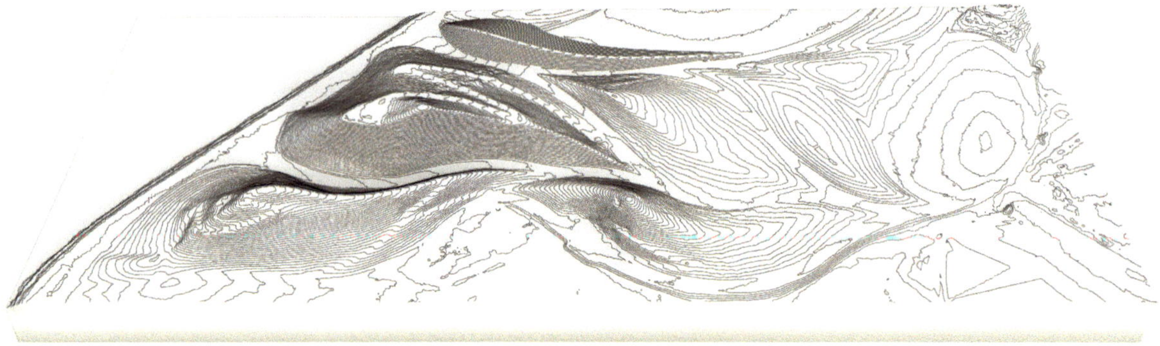

Cutting Plane Visualize sea level rise with a cutting plane. In Rhino model a rectangular surface at mean sea level as the cutting plane. Simply move the cutting plane up or down to visualize changes in sea level. Plug this surface into a `Move` component as the input geometry. Set the motion to the `Unit Z` vector and set its multiplication factor with a number slider.

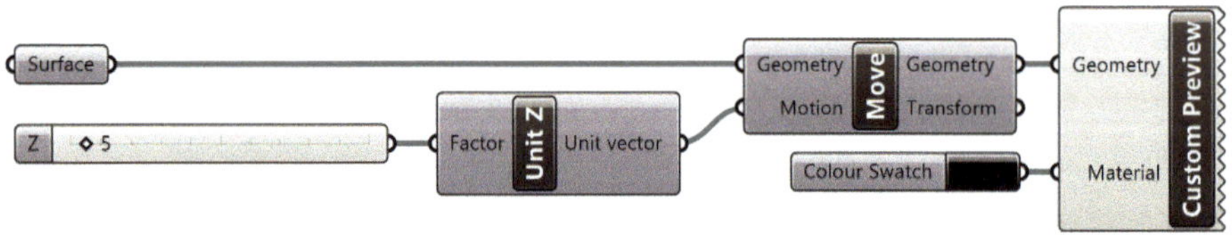

Export Shift the terrain back to its global coordinates and export it as a .xyz point cloud. Use the dfGridGlobal component to shift the edited distance field back to its original global coordinates. Then use the dfExportXYZ component to export the distance field as a .xyz point cloud. Set the name for the output dataset with a panel and use a Boolean Toggle to activate the component.

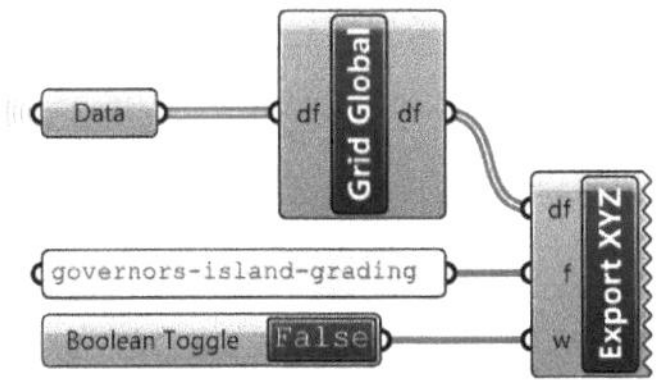

Gradient Apply a color gradient to the terrain mesh with the new landform. Deconstruct the mesh, remap its z-components to use as parameters for a color gradient, and then reconstruct the mesh with the new colors. First extract the vertices and faces from the terrain mesh with a Deconstruct Mesh component. Plug the vertices into a Deconstruct component to extract their z-values.

Find the domain of the z-values with a `Bounds` component and then remap the z-values to a domain from 0 to 1 with a `Remap Numbers` component. Construct a color gradient by plugging the list of z-values into a `Gradient` component as the input parameter. Finally reconstruct the mesh with a `Construct Mesh` component. Set the input vertices to the original vertices, the input faces to the original faces, and the input color to the gradient.

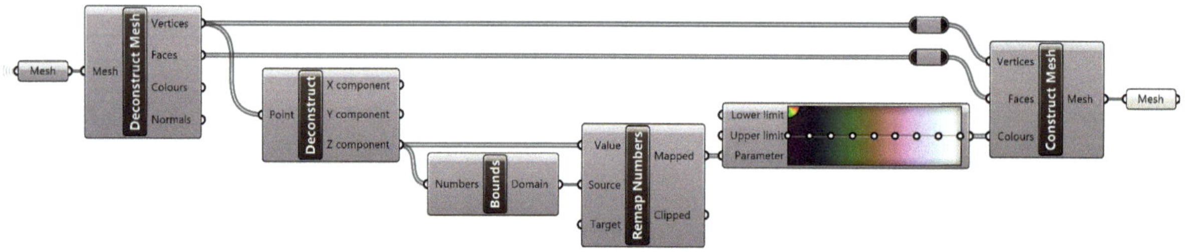

References

[1] [Software] Autodesk, *Civil3D*. url: https://www.autodesk.com/products/civil-3d.

[2] [Software] Vectworks, *Landmark*. url: https://www.vectorworks.net/.

[3] [Software] Landau Design+Technology, *Land Kit* version 1.2, 2022. url: https://www.landkit.design/, vcs: https://www.food4rhino.com/en/app/land-kit.

[4] Mathias Bernhard, Dana Cupkova, Fujan Fahmi, Christophe Girot, Fabio Gramazio, Francesca Hughes, Ilmar Hurkxkens, Marco Hutter, Dominic Jud, Matthias Kohler, Benedikt Kowalewski, Jesus Medina, and Ammar Mirjan. *Robotic Landscapes – Designing the Unfinished*. en. Edited by Ilmar Hurkxkens, Fujan Fahmi, and Ammar Mirjan. Zurich: Park Books, 2021. isbn: 978-3-03860-254-5.

[5] Dominic Jud, Ilmar Hurkxkens, Christophe Girot, and Marco Hutter. "Robotic embankment". In: *Construction Robotics* 5.2 (June 2021), pages 101–113. issn: 2509-8780. doi: 10.1007/s41693-021-00061-0.

[6] Ilmar Hurkxkens. "Robotic Landscapes: Topological Approaches to Terrain, Design, and Fabrication". PhD thesis. Zurich: ETH Zurich, 2020. doi: 10.3929/ethz-b-000451100.

[7] Ilmar Hurkxkens, Benedikt Kowalewski, and Christophe Girot. "Informing Topology: Performative Landscapes with Rapid Mass Movement Simulation". In: *Journal of Digital Landscape Architecture* 2020.5 (2020), pages 130–137. issn: 2367-4253. doi: 10.14627/537690014.

[8] Ilmar Hurkxkens and Mathias Bernhard. “Computational Terrain Modeling with Distance Functions for Large Scale Landscape Design”. In: *Journal of Digital Landscape Architecture* 2.4 (2019), pages 222–230. doi: 10.14627/537663024.

[9] [Software] Ilmar Hurkxkens and Mathias Bernhard, *Docofossor* version 0.904, 2019. lic: MIT License. vcs: https://github.com/dbt-ethz/docofossor.

[10] Mathias Bernhard, Michael Hansmeyer, and Benjamin Dillenburger. “Volumetric modelling for 3D printed architecture”. In: *AAG 2018: Advances in Architectural Geometry 2018*. Edited by Lars Hesselgren, Axel Kilian, Samar Malek, Karl-Gunnar Olsson, Olga Sorkine-Hornung, and Chris Williams. Wien: Klein Publishing, 2018, pages 392–415. isbn: 978-3-903015-13-5.

[11] [Software] Bison Development Team, *Bison* version 0.9, 2018. url: https://www.food4rhino.com/en/app/bison.

[12] Georges Descombes. *Aire: The River and Its Double*. Park Books, 2018. isbn: 9783038600817.

[13] Snøhetta. *MAX IV Laboratory Landscape*. 2016. url: https://snohetta.com/projects/70-max-iv-laboratory-landscape (visited on 07/28/2021).

[14] West 8. *Governor's Island: The Hills*. 2016. url: https://www.west8.com/projects/governors_island_phase_2_the_hills (visited on 03/05/2023).

[15] H+N+S. *Land Art Park Buitenschot*. 2013. url: https://www.hnsland.nl/en/projects/land-art-park-buitenschot (visited on 05/20/2023).

[16] Marcel J.F. Stive, Matthieu A. de Schipper, Arjen P. Luijendijk, Stefan G.J. Aarninkhof, Carola van Gelder-Maas, Jaap S.M. van Thiel de Vries, Sierd de Vries, Martijn Henriquez, Sarah Marx, and Roshanka Ranasinghe. “A New Alternative to Saving Our Beaches from Sea-Level Rise: The Sand Engine”. In: *Journal of Coastal Research* 29.5 (2013), pages 1001–1008. doi: 10.2112/JCOASTRES-D-13-00070.1.

[17] Peter Petschek. *Grading for Landscape Architects and Architects*. Birkhäuser, 2012. isbn: 978-3-0346-0987-6.

[18] Rijkswaterstaat. *The Sand Motor*. 2011. url: https://dezandmotor.nl/en (visited on 05/20/2023).

[19] Hargreaves and Jones. *South Pointe Park*. 2009. url: http://www.hargreaves.com/work/south-pointe-park (visited on 05/20/2023).

18 Fabrication

18.1 Digital Fabrication

Technologies for digital fabrication such as 3D printing and computer numerical control (CNC) machining are used by designers to build scale models and custom parts. Designers use digital fabrication to precisely reconstruct scale models of cities or landscapes from data such as airborne lidar. They also use digital fabrication to rapidly build scale models of computationally designed buildings or landforms. Digital fabrication is also used for non-standard construction, for building unique, custom elements [31, 25, 19]. Landscape architects digitally fabricate site models and custom elements such as fountains, benches, and paving. CNC machining was used, for example, to cut stonework for the Diana, Princess of Wales Memorial Fountain by Gustafson Porter + Bowman [32], the fountain for CityCenterDC by Gustafson Guthrie Nichol [22], the retaining walls for the National Museum of African American Culture and History by Gustafson Guthrie Nichol [17], and the paving, wall, and benches for Dilworth Park by Olin Studio [21]. Digital fabrication is also used to build works of public art such as the Faces of Elysian by Greenmeme [12].

This chapter covers the digital fabrication of models. Topics include CNC milled terrain, 3D printed terrain, and 3D printed trees. To learn more about digital fabrication read *Manufacturing Processes for Design Professionals* [29], *Digital Fabrication in Architecture* [23], *Digital Vernacular: Architectural Principles, Tools, and Processes* [20], *Material Strategies in Digital Fabrication* [15], and *Advanced 3D Printing with Grasshopper* [8].

18.1.1 Subtractive Processes

Subtractive technologies for digital fabrication create form by selectively removing material. These technologies include laser cutting and CNC machining processes such as milling, routing, water jet cutting, and plasma cutting. CNC machines follow a sequence of machine control instructions written in languages like G-code. Software for generating toolpaths for CNC machining include Fusion360, Mastercam, and RhinoCAM. With 3-axis CNC milling , designers can carve forms out of solid blocks of stock. Designers mill scale models of sites or their designs out of materials such as extruded polystyrene foam, medium-density fiberboard, or high-density urethane foam. Designers also mill non-standard components of their designs out of buildings materials such as wood and stone. CNC milling was used, for example, to cut spruce plywood panels for dECOi architects' One Main [26], to shape the wooden core of Snøhetta's Norwegian Wild Reindeer Centre Pavilion [28], and to carve sandstone blocks for the curving façade of Carl Fredrik Svenstedt Architect's Delas Frères Winery [10].

18.1.2 Additive Processes

3D printing is an additive process in which material is built up layer by layer. Common 3D printing processes include fused deposition modeling, stereolithography, selective laser sintering, binder jetting, ceramic extrusion, and concrete extrusion. Desktop 3D printers – using processes like fused deposition modeling – are often used by designers to fabricate scale models. In fused deposition modeling, the printer melts and extrudes plastic filament; as the extruded molten plastic rapidly cools, it fuses with the layer beneath. 3D printing is also used to autonomously construct buildings, bridges, and other structures. The White Tower by Benjamin Dillenburger and Michael Hansmeyer [5, 6] is an example of autonomous construction with extruded concrete printing, while examples of earth building with extruded mud printing include Pahoehoe Beauty by MAEID [13, 11], TECLA by Mario Cucinella Architects [7], and Casa Covida by Emerging Objects [9].

18.2 3D Printing Terrain

To 3D print terrain, the ground should be modeled as a solid volume, rather than a surface. Chapters 13, 15, 16, and 17 introduced methods for procedurally generating terrain, modeling terrain from point clouds, and grading earthworks. These methods generate meshes that represent terrain as a surface, rather than a volume. This section covers how to prepare terrain models for 3D printing by converting surface meshes into closed, solid volumes.

Code: `solidify.gh`

Input Mesh Begin by generating a terrain mesh with Grasshopper. This example uses a mesh generated from cellular noise with `fast-noise.gh`. Set your mesh as an input parameter.

Project Base Create a base for the 3D printed terrain. Start by deconstructing the terrain mesh to extract its vertices and faces. Then project the vertices onto a plane underneath the terrain. Reconstruct the mesh using the projected vertices and the original faces. To do this, plug your mesh into a `Deconstruct Mesh` component to extract the vertices and faces. Then plug the vertices into a `Project` component. Set the projection plane to the `Unit Z` vector and set its factor to a negative value on a number slider. Plug the projected vertices into a `Construct Mesh` component and set the faces to the original faces from the deconstructed mesh. This will generate a mesh base with the same number of vertices and edges as the terrain. The exterior vertices will be used to model the sides of the 3D print.

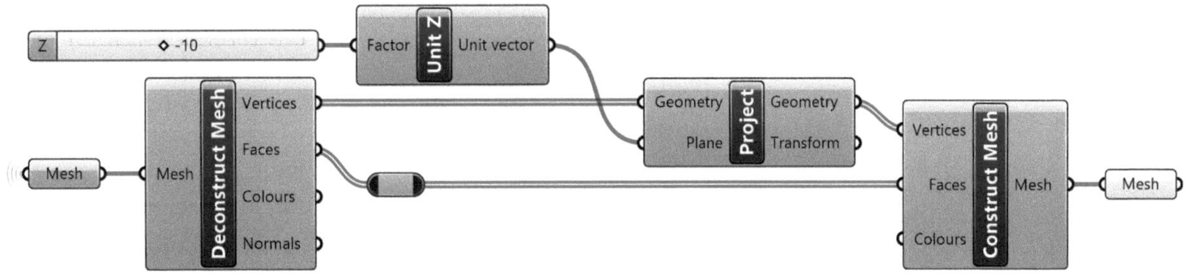

Construct Sides Model the sides of the 3D print from the exterior vertices of the terrain mesh and its base. Extract the edges of the terrain mesh with a `Mesh Edges` component. Flatten the naked edges – the edges on the exterior of the mesh – and plug these into an `End Points` component to extract their vertices. Do the same for the base mesh. Then add a `Merge` component. Plug the start points and end points of the terrain mesh into `Merge` as input data streams `D1` and `D2` respectively. Plug the start points of the base mesh into `Merge` as input data stream `D4`, while plugging the end points into input data stream `D3`. Graft all of these inputs. Plug the resulting data tree of points into a `Construct Mesh` component and flatten the output into a list of meshes.

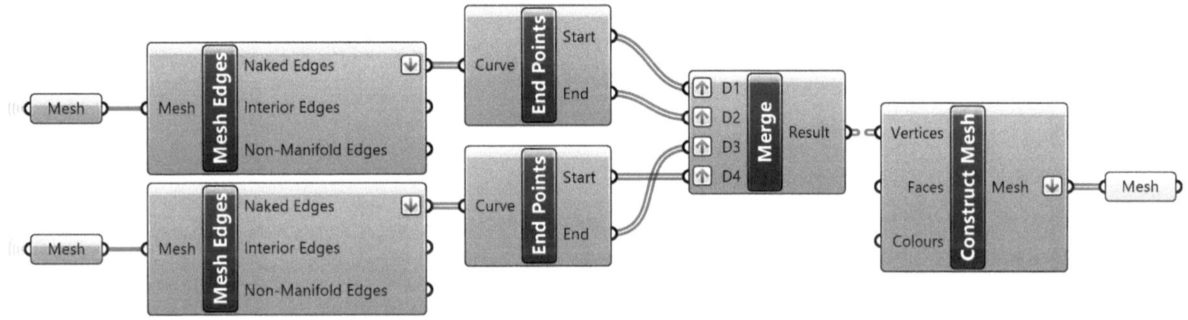

Join Meshes Join, unify, combine, clean, and bake the meshes before exporting for 3D printing. Plug the terrain, base, and side meshes into a `Mesh Join` component and flatten these inputs. Plug the resulting mesh into a `Unify Mesh` component so that the mesh normals all point outside of the mesh. A mesh normal is a direction vector perpendicular to the face of the mesh. Then plug the unified mesh into a `Combine&Clean` component. Finally bake the mesh and export it as either a `.stl` or `.3mf` file for 3D printing.

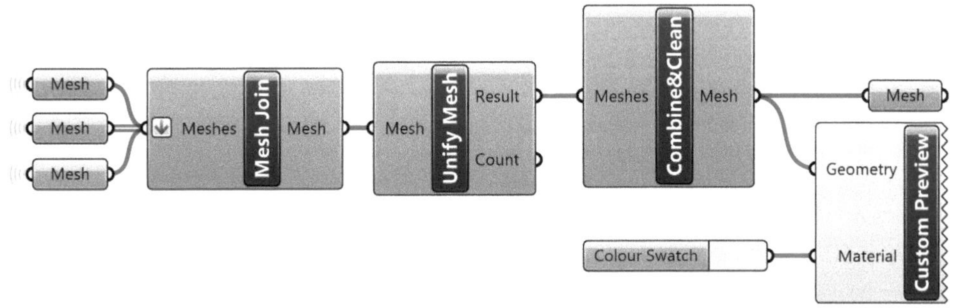

18.3 3D Printing Trees

Use volumetric modeling to prepare a point cloud model of a tree for 3D printing. Trees are challenging to 3D print because of their fractal branching structure. They have a multitude of tiny, overhanging parts – of branches and leaves – that may require support structure and may not print well at a given scale, depending on the printing technology. 3D printing processes such as fused deposition modeling build temporary support structures for overhanging parts. Printing a tree with these technologies requires such an excessive number of supports to build the branches that it is impractical to remove it all. Trees, however, can be 3D printed without support structures using technologies such as binder jet 3D printing or selective laser sintering that build the model in a bed of powder.

Point clouds can be modeled volumetrically as voxels. Voxels are 3-dimensional cells in a regular grid, i.e. a 3-dimensional raster. To voxelize a point cloud, first construct an empty 3-dimensional grid. If at least one point in the point cloud falls within a grid cell, then that cell is occupied. The occupied cells – the voxels – form a volume. The detail of this volume depends upon the resolution of the grid. Volumetric modeling is useful for 3D printing point cloud models of complex things like trees. While a fine resolution grid of voxels may generate gaps and parts that are too thin to print, a coarser resolution grid will fill gaps and thicken these parts. Meshes for 3D printing can be generated from voxels using algorithms such as marching cubes [34], ball pivoting [33], or Poisson surface reconstruction [30]. Plugins for volumetric modeling in Grasshopper include `Volvox` [16, 18], `Dendro`, and `VoxelTools` [3]. The Python library `Open3D` [14, 2] also has methods for voxelization and surface reconstruction.

18.3.1 Volumetric Modeling in Grasshopper

In Grasshopper, use the plugin `Dendro` to construct a voxel model of a tree from a laser-scanned point cloud. Dendro is a plugin for volumetric modeling that is built on the `OpenVDB` library, which was developed by Dreamworks and is now maintained by the Academy Software Foundation. Dendro has components for converting geometry into volumes and calculating Boolean operations. Use `voxelize.gh` to model a tree or use `voxelize-tree.gh` to model a tree with a base.

Figure 18.1: Voxel size

Plugin: `Dendro`

Data: `laurus-nobilis-01-s.3dm`

Code: `voxelize.gh`

Input Open, scale, and explode a point cloud model of a tree. Open a point cloud such as `laurus-nobilis-01-s.3dm` in Rhino. All of the models in the `Cloud Forest` library of laser-scanned trees are in meters. Since millimeters will be more useful for voxelization and 3D printing, change the units in Rhino and scale from meters to millimeters. Run the `Units` command in Rhino's command line and change the model units. Changing the units will trigger a prompt about automatic scaling. In Rhino 8, choose to change the unit system and maintain object sizes. In older versions of Rhino, agree to scale the model by 1000. The original model of the Bay Laurel is almost 4 meters tall, so the rescaled model should be almost 4000 millimeters tall. Because Dendro works with points rather point clouds, explode the point cloud with the `Explode` command. Select all of the points and set them as a point parameter in Grasshopper.

```
Units
Explode
```

Settings First define the settings for the voxel model. Add Dendro's `Create Settings` component and set the voxel size, isovalue, and adaptivity parameters. Voxel size is the key parameter; it defines the density of the 3D grid and the resolution of the volumetric model (Figure 18.1). While smaller voxels will yield a more detailed model, larger voxels will fill more gaps connecting more clusters of points. Use a relatively large voxel size so that clusters of points merge together to form a single volume. Dendro's default isovalue of zero may result in invalid meshes, so try setting very small isovalues. High adaptivity will result in fewer polygons, but may also generate unprintable, open meshes, so try a low setting.

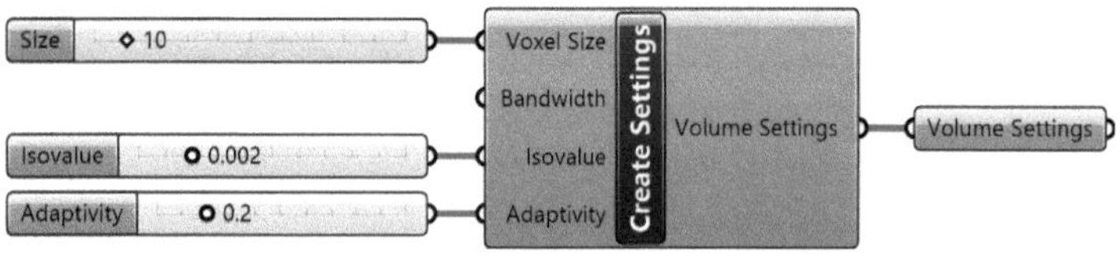

Voxelization Dendro generates volumes from points by modeling a sphere of voxels around each point. Use Dendro's `Points to Volume` component to create a voxel model from points. Set the point input to the exploded point cloud. Set the point radius large enough so that the resulting volumes merge together. Use the volume settings from the previous step.

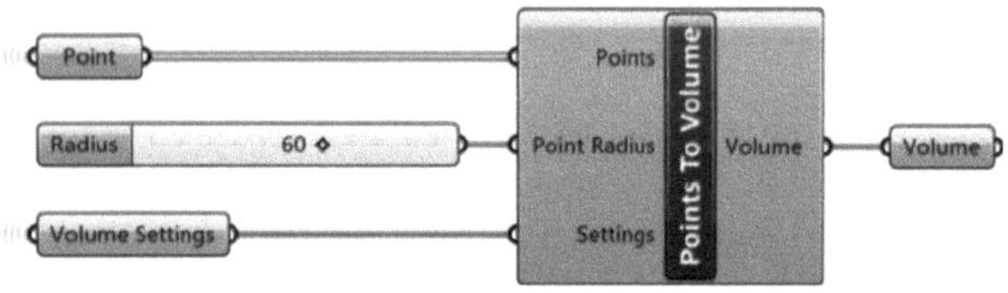

Smoothing Smoothing the voxel model of the tree will blend the connections between the spheres, making it more printable. Use Dendro's `Smooth Volume` component to smooth the model. Set the smoothing type to 1 for Laplacian smoothing.

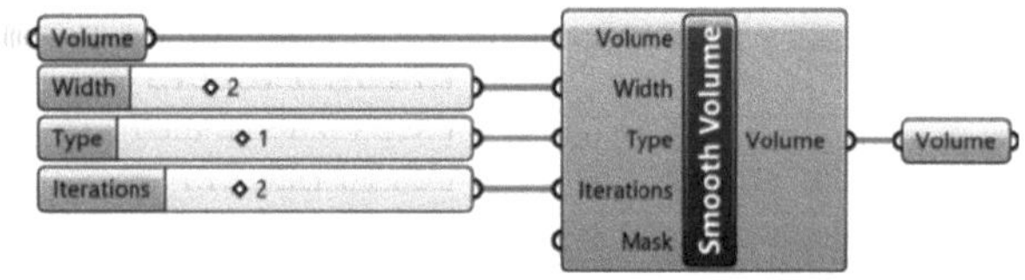

Meshing Convert the volume to a mesh with a `Volume to Mesh` component. Use the same volume settings.

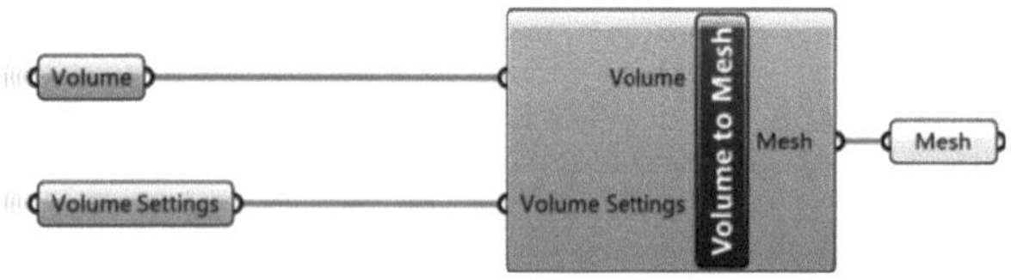

Culling Voxelizing a point cloud may result in many separate volumes. While increasing the voxel size and the point radius may merge many of the volumes together, some outliers may need to be removed. After converting the volume into a mesh, use the `Disjoint Mesh` component to isolate disconnected meshes. Cull all but the largest of these. To do this, first use a `Deconstruct Mesh` component to extract the faces from each of these meshes. Calculate the number of faces in each mesh with a `List Length` component and flatten the result. Then use a `Larger Than` component with a `Cull Pattern` to filter out the smaller meshes. Bake, scale, and export the remaining mesh as a `.stl` or `.3mf` file for 3D printing.

```
Scale
Export
```

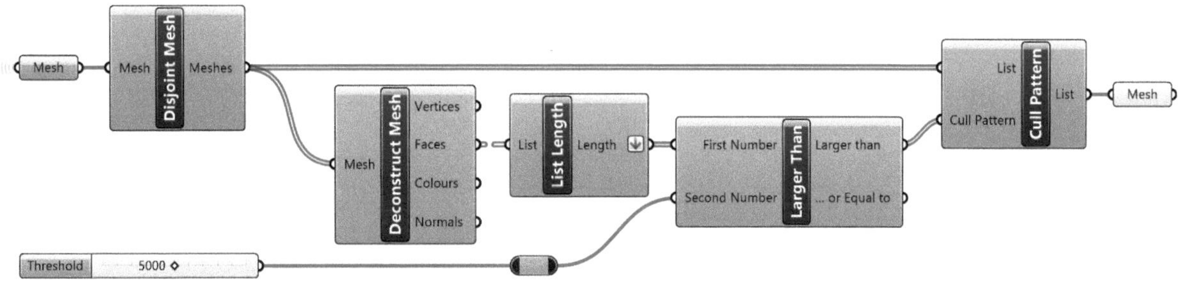

Base Consider adding a base beneath the tree so that the 3D printed model can stand freely. Use `voxelize-tree.gh` as a reference. To add a base, first model the base, convert it to a volume, merge it with the model of the tree using Dendro's `Volume Union` component, smooth the resulting volume, and convert it to a mesh.

Code: `voxelize-tree.gh`

Figure 18.2: Voxelized tree

18.3.2 Marching Cubes in Grasshopper

In Grasshopper =, use the plugin `VoxelTools` to construct a voxel model of a tree from a laser-scanned point cloud. VoxelTools is a plugin for volumetric modeling with components for voxelizing geometry, calculating Boolean operations, and generating meshes with the marching cubes algorithm. It has been used for projects such as The Why Factory's Porocity [24]. Use `voxelate.gh` to model a tree or use `voxelate-tree.gh` to model a tree with a base.

Plugin: `VoxelTools`

Data: `laurus-nobilis-01-s.3dm`

Code: `voxelate.gh`

Input Open, scale, and explode a point cloud model of a tree. First open a point cloud such as `laurus-nobilis-01-s.3dm` in Rhino. Run the `Units` command in Rhino's command line and change the model units and automatically scale the point cloud. Then explode the point cloud with the `Explode` command. Select all of the points and set them as a point parameter in Grasshopper.

```
Units
Explode
```

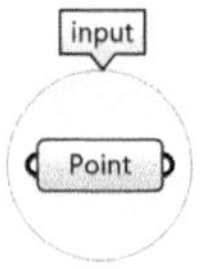

Voxelate Start by constructing an empty 3D grid of voxels inside of a bounding box. Create a box just larger than the point cloud to use as the bounding box. Then use VoxelTool's `Create VoxelGrid` component to generate the empty voxel grid. Set the bounding box to the box and set the voxel size of the grid in the x-, y-, and z-axes with number sliders. Populate the voxel grid from the point cloud with VoxelTool's `VoxelateGeometry` component. Set the input geometry to the list of points from the exploded point cloud, set the voxel grid, and set the distance with a number slider.

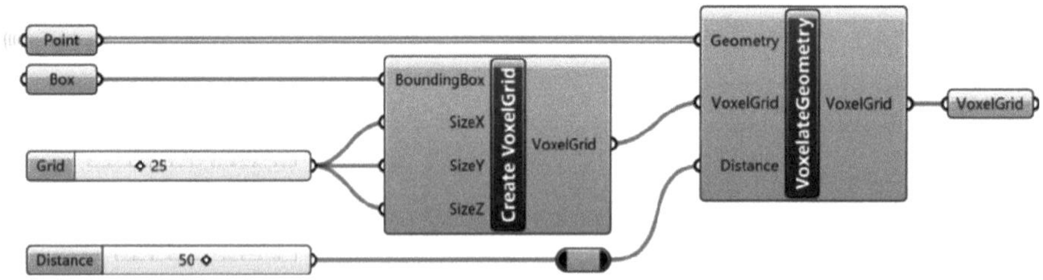

Marching Cubes Convert the voxel grid into a mesh with VoxelTool's `Hull from Marching Cubes` component. Set the isolevel parameter with a number slider.

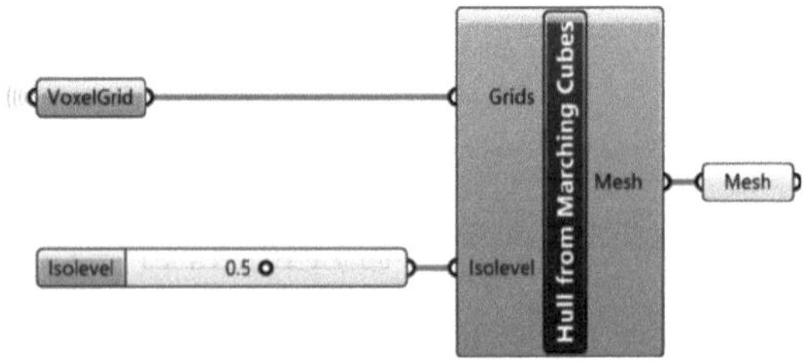

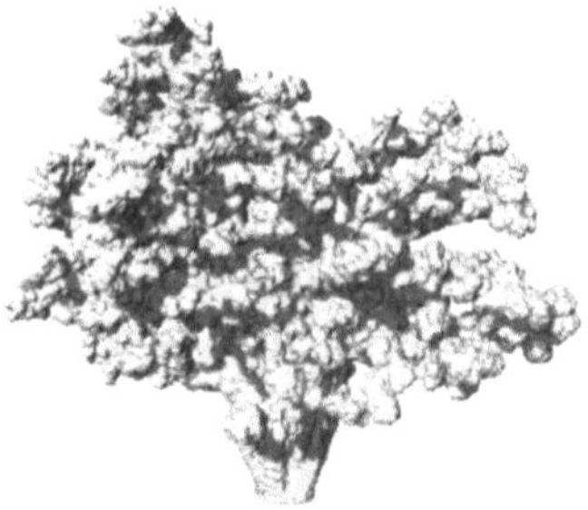

Culling Since voxelizing a point cloud can result in many separate volumes, cull the outliers so that only the largest remains. Then bake, scale, and export the remaining mesh.

```
Scale
Export
```

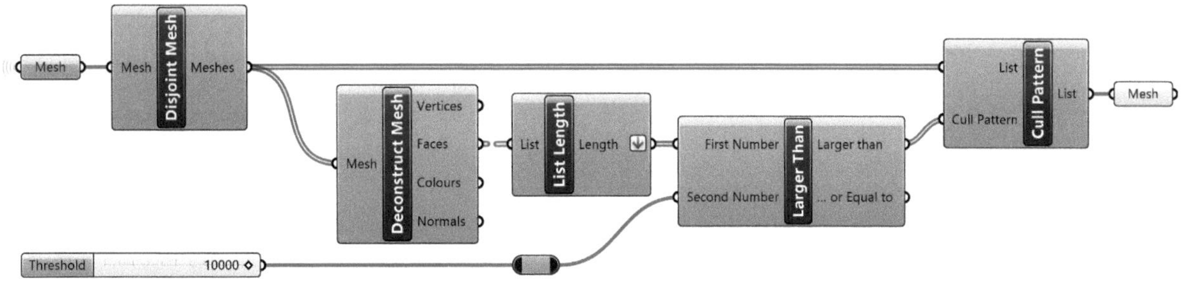

Base Use `voxelate-tree.gh` to add a base underneath the tree. To add a base, first model the base, convert it to a voxel grid, merge it with the model of the tree using VoxelTool's `VoxelGrid Boolean Union` component, and then convert it to a mesh using the marching cubes algorithm.

Code: `voxelate-tree.gh`

Figure 18.3: Surface reconstruction of a tree with the marching cubes algorithm

18.3.3 Volumetric Modeling with Python

Voxelize a laser-scanned tree using Python with the Open3D library. `Open3D` is a Python library for processing and visualizing 3D geometry [14, 2]. It supports data structures including images, point clouds, voxels, and meshes. Use Open3D to import and voxelize a point cloud, render the voxels, convert them into a mesh, and then export the mesh for 3D printing.

Data: `laurus-nobilis-01-l.pcd`

Code: `voxelize.py`

Figure 18.4: Voxel sizes

Import First import open3D as the alias o3d for the sake of brevity. Then import the point cloud laurus-nobilis-01-l.pcd with the function open3d.io.read_point_cloud. Open3D can read and write point clouds in .xyz, .ply, .pts, or .pcd format. While Open3D only supports colorized xyzrgb ASCII point clouds with floating point colors between 0 and 1, many point clouds – including those in the Cloud Forest library – use integer colors between 0 to 255. For this example use the Point Cloud Library's .pcd format [27].

Voxelize Voxelize the point cloud with the method open3d.geometry.VoxelGrid.create_from_point_cloud. Then render the voxels with the function open3d.visualization.draw_geometries. An interactive rendering will open in another window; use the mouse to navigate and press p to capture a screenshot or h to print a list of commands. Adjust the size of the voxels and rerun the script until the voxels coalesce into a single, printable volume.

Meshing To convert the voxels into a mesh, loop through the voxels and model each one as a mesh box. First create an empty triangular mesh with the function open3d.geometry.TriangleMesh. Then get the voxels from the voxel grid with the method open3d.geometry.VoxelGrid.get_voxels and use a **for** loop to iterate through the list of voxels. For each voxel, create a mesh box with the method open3d.geometry.TriangleMesh.create_box, assign it a color based on its voxel with open3d.geometry.TriangleMesh.paint_uniform_color, move it to the voxel's position with open3d.geometry.TriangleMesh.translate, and then add it to the mesh.

Export Export the mesh as a .stl for 3D printing. Calculate normals for the mesh with the function open3d.geometry.TriangleMesh.compute_triangle_normals. Then use the function open3d.io.write_triangle_mesh to export the mesh.

```
# import modules
import open3d as o3d

# set path
data = 'laurus-nobilis-01-l.pcd'

# read point cloud
cloud = o3d.io.read_point_cloud(data, format='pcd')

# voxelize
size = 0.05
volume = o3d.geometry.VoxelGrid.create_from_point_cloud(
    cloud,
    voxel_size=size
    )
```

```python
# render voxels
o3d.visualization.draw_geometries([volume])

# convert voxels to mesh
mesh = o3d.geometry.TriangleMesh()
voxels = volume.get_voxels()
for voxel in voxels:
    box = o3d.geometry.TriangleMesh.create_box()
    box.paint_uniform_color(voxel.color)
    box.translate(voxel.grid_index)
    mesh += box

# export mesh
o3d.io.write_triangle_mesh(
    'voxels.stl',
    mesh.compute_triangle_normals()
    )
```

Figure 18.5: Voxelized tree

18.4 CNC Milling Terrain

CNC mill a model of the Hills of Governors Island in high-density foam using RhinoCAM [1]. The model shown in this book was milled out of a two inch thick block of high-density urethane foam, using carbide end mills as cutting tools. High-density urethane foam boards are a good material for model making. The products Renshape 5020 and Signfoam are relatively light and mill easily with fine detail. For additional guidance on CNC milling with RhinoCAM, read the RhinoCAM manual [4] or follow my tutorial at https://youtu.be/cBSGlFYaROw.

Data: `governors-island-cnc.3dm` & `cnc-tools.csv`

Plugin After installing `RhinoCAM`, restart Rhino and set the units to inches. In the RhinoCAM menu, select the Mill module. This will open the RhinoCAM Machining Browser panel for programming and simulating toolpaths and the Machining Objects panel for creating tools.

Model Start with a terrain mesh. Either generate and bake a terrain mesh with `grading.gh` or open `governors-island-cnc.3dm`. Rescale the terrain mesh from its real-world size to a model scale. In Rhino's command line run the `Scale` command, set the base point to either automatic or the origin at `0,0,0`, and set the scale factor to `0.01`.

```
Scale
0,0,0
0.01
```

Try vertically exaggerating the model. In the command line, run the `Scale1D` command, set the base point to the origin at `0,0,0`, set the scale factor to `2`, and set the scale direction to the vector `0,0,1`.

```
Scale1D
0,0,0
2
0,0,1
```

Tools The next step is to define milling tools in RhinoCAM's Machining Objects panel. Use quarter and eighth inch diameter carbide ball nosed end mills. Either select `Load Tool Library` to import `cnc-tools.csv` or select `Create/Edit Tools` to define the tools described in table 18.1. Tools definitions with feeds and speeds for a given material can be saved as `.csv` text file.

Table 18.1: Tools

Diameter	Speed	Plunge	Approach	Engage	Cut	Retract	Departure
0.25	15000	20	180	180	200	180	180
0.125	18000	20	75	175	150	75	175

Machine Define the settings for your CNC machine. The settings will depend on the machine in your fabrication facilities. In the machine setup tab of RhinoCAM's Machining Browser, select `Machine` and set the machine type to 3-axis. In `Post` set the post-processor and the posted file extension for the G-code. Our Fabrication Lab, for example, uses Velocity CNC as the post-processor for our Forest Scientific CNC machine with `.nc` files for G-code.

Stock Define the stock – a two inch thick block of urethane foam – and align it with the model. For this example, the stock was already cut to match the width and length of the scale model. In the program tab of RhinoCAM's Machining Browser, in the `Stock` dropdown menu, pick `Box Stock`. Set the dimensions of the stock to `7.5 x 17 x 2` inches. Then in the `Align` dropdown menu, pick `Align Stock`. Set the z-alignment to the top of the stock. This will reduce the amount of milling and provide a generous base for the model. In the `Align` dropdown menu, select `Set WCS` to set the world coordinate system to a top corner. Set the zero face to the highest z value and set the zero position to the southwest corner. This will be the work zero, the origin of the program. Later, you will need to set the program zero for the milling machine to match this datum.

Machining Operations Generate a parallel finishing toolpath with the quarter inch diameter end mill. The CNC milling machine will move the tool, which is spinning 15,000 rotations per minute, in parallel passes through the stock to excavate the model. As the tool moves in passes along the x-axis, stepping over row by row on the y-axis, it will move up and down on the z-axis, following the elevation of the mesh. Each pass will be offset by 25 percent of the tool's diameter. Be sure that the flute length of your tool is long enough for the deepest cut in this program. If the tool's flutes are not long enough, use a rough cut to remove excess material from the stock before running a finishing operation.

In the Machining Operations panel in the `3-Axis Advanced` dropdown menu, select `Parallel Finishing` to generate a toolpath. In the `Parallel Finishing` dialog, open the `Tool` tab and select the `HDU 0.25" 2FL Ball Upcut` tool. In the `Feeds and Speeds` tab, load the settings from the tool. In the `Cut Parameters` tab, set the stepover to 25 percent of the tool diameter. Finally, generate the toolpath. This operation will mill a nice terrain model with a relatively smooth surface, but there will be noticeable grooves left by the tool along the x-axis (Figure 18.7). To mill a smoother surface, run another parallel finishing operation along the y-axis with a smaller eighth inch end mill and thus a smaller stepover. Create another parallel finishing toolpath. Set the tool to `HDU 0.125" 2FL Ball Upcut` and load the feeds and speeds from the tool. In the `Cut Parameters` tab, set the stepover to 25 percent of the tool diameter and set the angle to 90 degrees. Then generate the toolpath.

Simulation Simulate both toolpaths to preview the process and check for collisions. Running simulations is critical for making beautiful models and avoiding mishaps. In the simulate tab of RhinoCAM's Machining Browser, press play to start a simulation (Figure 18.6). Once the simulation is complete, right click on the toolpath in the machining job browser and post it as a `.nc` G-code file. Load this G-code file on the computer that controls the CNC machine to start the operation.

Milling Once your program is ready, prepare the stock and the machine. First, align your stock on the machine bed and secure it so that it will not move while being milled. Workholding methods include tabs, clamps, or double-sided tape. For this example we used Nitto P-02, a double-sided machining tape. Load your G-code in the machine control program for the CNC machine. Then set the program zero by moving the tool to the top southwest corner of your stock. Don ear and eye protection, turn on the spindle, start the vacuum, and then run the program.

Figure 18.6: Simulation

Figure 18.7: Parallel finishing with a quarter inch diameter tool

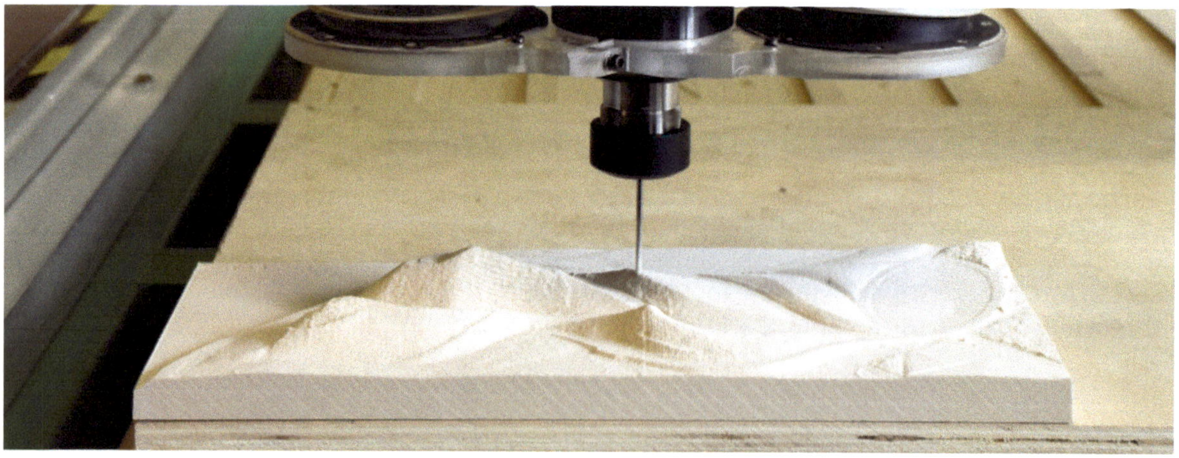

Figure 18.8: Parallel finishing with an eighth inch diameter tool

References

[1] [Software] MecSoft, *RhinoCAM* version 2023, 2023. url: https://mecsoft.com/products/rhinocam/.

[2] [Software] Open3D Development Team, *Open3D: A Modern Library for 3D Data Processing* version 0.17, 2023. lic: MIT License. url: http://www.open3d.org/, vcs: https://github.com/isl-org/Open3D.

[3] [Software] Arend van Waart, *Voxeltools: Lightweight voxel geometry and Boolean voxel operations for Grasshopper* version 1.03, 2023. lic: MIT License. url: https://www.food4rhino.com/en/app/voxeltools, vcs: https://github.com/arendvw/voxeltools.

[4] MecSoft. *RhinoCAM-MILL 2022 Quick Start Guide*. 2022. url: https://mecsoft.com/QuickStartGuides/RhinoCAM2022/MILLQuickStartGuide.pdf.

[5] Benjamin Dillenburger. *Tor Alva*. 2021. url: https://dbt.arch.ethz.ch/project/tor-alva/ (visited on 06/25/2022).

[6] Michael Hansmeyer. *White Tower*. 2021. url: https://www.michael-hansmeyer.com/white-tower (visited on 06/25/2022).

[7] Mario Cucinella Architects. *TECLA*. 2020. url: https://www.mcarchitects.it/en/projects/tecla-technology-and-clay (visited on 06/25/2022).

[8] Diego Garcia Cuevas and Gianluca Pugliese. *Advanced 3D Printing with Grasshopper: Clay and FDM*. 2020. isbn: 979-8635379011.

[9] Emerging Objects. *Casa Covida*. 2020. url: https://emergingobjects.com/project/casa-covida/ (visited on 06/25/2022).

[10] Carl Fredrik Svenstedt Architect. *Delas Frères Winery*. 2019. url: https://www.carlfredriksvenstedt.com/what/15722322/ (visited on 06/25/2022).

[11] Daniela Mitterberger and Tiziano Derme. "Soil 3D Printing". In: *ACADIA – Ubiquity and Autonomy: Proceedings of the 39th Annual Conference of the Association for Computer Aided Design in Architecture* (2019), pages 586–595. doi: https://doi.org/10.52842/conf.acadia.2019.586.

[12] Greenmeme. *Faces of Elysian Valley*. 2018. url: https://www.greenmeme.com/art/specialists-tfhkh-gf2a9 (visited on 06/24/2023).

[13] MAEID. *Pahoehoe Beauty*. 2018. url: https://maeid.com/works/pahoehoe-beauty (visited on 06/25/2022).

[14] Qian-Yi Zhou, Jaesik Park, and Vladlen Koltun. "Open3D: A Modern Library for 3D Data Processing". In: *arXiv:1801.09847* (2018). doi: 10.48550/arXiv.1801.09847.

[15] Christopher Beorkrem. *Material Strategies in Digital Fabrication*. Routledge, 2017. isbn: 978-1138654204.

[16] [Software] Henrik Leander Evers and Mateusz Zwierzycki, *Volvox* version 0.3.0.0, 2016. Centre for Information Technology and Architecture. url: https://www.food4rhino.com/en/app/volvox, vcs: https://github.com/CITA-cph/Volvox/tree/master.

[17] Gustafson Guthrie Nichol. *National Museum of African American History and Culture*. 2016. url: https://www.ggnltd.com/national-museum-of-african-american-history-and-culture (visited on 06/24/2023).

[18] Mateusz Zwierzycki, Henrik Leander Evers, and Martin Tamke. "Parametric architectural design with point-clouds – Volvox". In: *Proceedings of the 34th eCAADe Conference*. Volume 2. University of Oulu, Oulu, Finland, 2016, pages 673–682. doi: 10.52842/conf.ecaade.2016.2.673.

[19] Achim Menges. *Material Synthesis: Fusing the Physical and the Computational.* Architectural Design. Wiley, 2015. isbn: 9781118878378.

[20] James Stevens and Ralph Nelson. *Digital Vernacular: Architectural Principles, Tools, and Processes.* EBL-Schweitzer. Taylor & Francis, 2015. isbn: 9781317548430.

[21] Olin Studio. *Dilworth Park.* 2014. url: https://www.theolinstudio.com/dilworth-park (visited on 06/24/2023).

[22] Gustafson Guthrie Nichol. *CityCenterDC.* 2013. url: https://www.ggnltd.com/citycenterdc (visited on 06/24/2023).

[23] Nick Dunn. *Digital Fabrication in Architecture.* Laurence King Publishing, 2012. isbn: 978-1-85669-891-7.

[24] The Why Factory. *Porocity.* 2012. url: https://thewhyfactory.com/project/porous-city (visited on 06/22/2023).

[25] Mario Carpo. *The Alphabet and the Algorithm.* Cambridge, MA: MIT Press, 2011, page 169. isbn: 978-0262515801.

[26] dECOi Architects. *One Main.* 2011. url: http://www.decoi-architects.org/2011/10/onemain/ (visited on 07/28/2021).

[27] Radu Bogdan Rusu and Steve Cousins. "3D is here: Point Cloud Library (PCL)". In: *IEEE International Conference on Robotics and Automation (ICRA).* Shanghai, China: IEEE, May 2011.

[28] Snøhetta. *Tverrfjellhytta, Norwegian Wild Reindeer Pavilion.* 2011. url: https://snohetta.com/projects/2-tverrfjellhytta-norwegian-wild-reindeer-pavilion (visited on 07/28/2021).

[29] Rob Thompson. *Manufacturing Processes for Design Professionals.* Thames & Hudson, 2007. isbn: 9780500513750.

[30] Michael Kazhdan, Matthew Bolitho, and Hugues Hoppe. "Poisson Surface Reconstruction". In: *Symposium on Geometry Processing.* Edited by Alla Sheffer and Konrad Polthier. The Eurographics Association, 2006. isbn: 3-905673-24-X. doi: 10.2312/SGP/SGP06/061-070.

[31] Mario Carpo. "Tempest in a Teapot". In: *Log* 6 (2005), pages 99–106. issn: 15474690. url: http://www.jstor.org/stable/41765067.

[32] Gustafson Porter + Bowman. *Diana, Princess of Wales Memorial Fountain.* 2004. url: https://www.gp-b.com/diana-princess-of-wales-memorial (visited on 06/24/2023).

[33] Fausto Bernardini, Joshua Mittleman, Holly Rushmeier, Cláudio Silva, and Gabriel Taubin. "The Ball-Pivoting Algorithm for Surface Reconstruction". In: *IEEE Transactions on Visualization and Computer Graphics* 5.4 (Oct. 1999), pages 349–359. issn: 1077-2626. doi: 10.1109/2945.817351.

[34] William E. Lorensen and Harvey E. Cline. "Marching Cubes: A High Resolution 3D Surface Construction Algorithm". In: *SIGGRAPH Computer Graphics* 21.4 (Aug. 1987), pages 163–169. issn: 0097-8930. doi: 10.1145/37402.37422.

A Packages

A.1 Grasshopper Plugins

Plugins for Grasshopper may be distributed as Grasshopper assemblies, binary installers, or packages. Most plugins are available from McNeel Europe's Food4Rhino repository, but some are only available from the plugin developer's website or repository. Watch this video to learn how to install plugins for Grasshopper. This book uses the plugins listed in Table A.1. Other recommended plugins are listed in Table A.2. Volvox is needed for Rhino 7 since it adds support in Grasshopper for point clouds, but is not needed for Rhino 8, which has a native point cloud parameter. SnappingGecko is optional, but highly recommended as it is an invaluable aid for neatly arranging definitions.

Table A.1: Required plugins

Plugin	Type	Link	
Bitmap+	Package	https://www.food4rhino.com/en/app/bitmap	[5]
Color+	Package	https://www.food4rhino.com/en/app/color	[6]
Dendro	Assembly	https://www.food4rhino.com/en/app/dendro	[15]
Docofossor	Assembly	https://www.food4rhino.com/en/app/docofossor	[23]
SnappingGecko	Package	https://www.food4rhino.com/en/app/snappinggecko	[20]
Volvox	Installer	https://www.food4rhino.com/en/app/volvox	[26]
Voxeltools	Assembly	https://www.food4rhino.com/en/app/voxeltools	[11]

Food4Rhino Most plugins are available from McNeel Europe's repository hosted at https://www.food4rhino.com. To download a plugin from this repository, register for a free account and then login. Download the latest release of the plugin. Windows may automatically block downloaded files from running, so right click on the downloaded file, open properties, and unblock it: Properties > Security > Unblock. If the download is a `.zip` archive, unblock and then unarchive it so that all of the contents are unblocked. Then move the unblocked and unarchived plugin to Grasshopper's Library directory. To find this library, open Grasshopper and then browse to File > Special Folders > Components. Move the plugin into this directory and restart Rhino and Grasshopper. If the plugin has a binary installer rather than a Grasshopper assembly, run the installer after unblocking it. Simple plugins shipped as Grasshopper assemblies should work on Mac, but plugins with `.exe` installers or `.dll` dynamic link libraries will not.

Package Manager Newer plugins will often be installed with Rhino's package manager [22]. This is a much easier way to install plugins. Plugins that are published with the package manager have an install button, rather than a download button on Food4Rhino. Click install to automatically open Rhino and the package manager. Once the package manager opens, just click install. These plugins can also be installed directly from Rhino. In the command line, run the command `PackageManager`. Then in the package manager, search for the name of the plugin and then click install.

PackageManager

Table A.2: Recommended plugins

Category	Plugin	Link	
Point Clouds	Cockroach	https://www.food4rhino.com/en/app/cockroach	[19]
	Tarsier	https://www.food4rhino.com/en/app/tarsier	[24]
	Volvox	https://www.food4rhino.com/en/app/volvox	[26]
Voxels	Dendro	https://www.food4rhino.com/en/app/dendro	[15]
	Voxeltools	https://www.food4rhino.com/en/app/voxeltools	[11]
Noise	Bitmap+	https://www.food4rhino.com/en/app/bitmap	[5]
	Noise4D	https://www.food4rhino.com/en/app/4d-noise	[28]
Terrain	Bison	https://www.food4rhino.com/en/app/bison	[25]
	Docofossor	https://www.food4rhino.com/en/app/docofossor	[23]
Landscape	Groundhog	https://www.food4rhino.com/en/app/groundhog	[17]
	Landkit	https://www.food4rhino.com/en/app/land-kit	[14]
Geospatial	Caribou	https://www.food4rhino.com/en/app/caribou	[13]
	Elk	https://www.food4rhino.com/en/app/elk	[27]
	Heron	https://www.food4rhino.com/en/app/heron	[16]
Robotics	KUKA|prc	https://robotsinarchitecture.org/	[12]
	Machina	https://www.food4rhino.com/en/app/machina	[21]
	Robots	https://www.food4rhino.com/en/app/robots	[10]
Utilities	Color+	https://www.food4rhino.com/en/app/color	[6]
	EleFront	https://www.food4rhino.com/en/app/elefront	[7]
	SnappingGecko	https://www.food4rhino.com/en/app/snappinggecko	[20]

A.2 Python Packages

Python software is distributed as packages. Package managers such as Pip and Conda are used to install, update, and uninstall packages. Because of the focus of this book is on design rather than development, a lot of packages are used. Rather than developing algorithms and code from scratch, the examples in this book build on software from the rich and ever growing ecosystem of free and open source software for scientific programming. Because this book uses a lot of packages, Conda is used to manage these packages. Software packages often have dependencies, i.e. other software on which they rely. Package managers will install all needed dependencies and try to resolve conflicts between dependencies. These conflicts can become a serious issue. Environments are used to minimize conflict between dependencies. An environment is an isolated collection of packages that will not conflict with other environments. This book uses a minimal distribution of Python called Miniconda, a Conda environment for developing and testing code, and another Conda environment for installing packages.

Conda This book uses `Conda` [2] as a package manager for Python. Conda can be used to install and update Python and useful packages like NumPy [4] and Scipy [8]. It can also create and manage environments. `Miniconda` [1] – a minimal Python distribution – is an easy way to install this package manager with the latest stable version of Python. Download and install Miniconda for your operating system from https://docs.conda.io/en/latest/miniconda.html. On Windows, the installer creates a shortcut for `Anaconda Prompt (miniconda3)`. Use this command line to run Conda commands on Windows. On Mac and Linux, just open your `terminal` and use Conda commands.

Spyder Install `Spyder` [3] – an integrated development environment for scientific computing with Python – in a dedicated environment using Conda. Creating a separate environment for Spyder reduces the risk of conflicting dependencies with other modules. Alternatively, use the binary installer for Spyder available for download at: https://www.spyder-ide.org

To install Spyder in a new environment, first open your terminal or command line interface. Initialize and update Conda. In Conda packages are distributed over different channels. Spyder and most of the packages used in this book are distributed over the `conda-forge` channel. Configure Conda to use only high priority channels, create a new environment called `spyder-env`, set `conda-forge` as the top priority channel, and then install Spyder and the other packages needed for developing and testing code. After activating this environment, launch Spyder with the command `spyder`. To leave an environment, either use the command `conda deactivate` or simply close the command line.

```
conda init bash
conda update -n base conda
conda config --env --set channel_priority strict
conda create -c conda-forge -n spyder-env spyder numpy scipy pandas matplotlib
  ↪ sympy cython
conda activate spyder-env
spyder
```

Computational Design Install the packages needed for this book in a new environment called `computational-design`. Most of the packages needed are distributed by `conda-forge`, but some will need to be installed with `pip`. After creating the environment and installing the packages available from `conda-forge`, activate the environment. Then install the packages that are only distributed via Pip.

To use this environment with Spyder, you will need to register it as Spyder's Python Interpreter. First activate the `computational-design` environment. Find the path for this environment using `where python` on Windows or `which python` on Mac or Linux. Deactivate this environment, activate `spyder-env`, and run Spyder. In Spyder, go to Tools > Preferences > Python Interpreter and set the interpreter to the path for the `computational-design` environment. Restart Spyder to use this environment.

```
conda create -n computational-design -c conda-forge spyder-kernels scipy seaborn
  ↪ rasterio lazrs-python laspy pyntcloud pyvista viscm colorspacious cmcrameri
  ↪ cmasher colorcet cmocean
conda activate computational-design
pip install lidario pyfastnoisesimd cellpylib
where python
conda deactivate
conda activate spyder-env
spyder
```

Always install Pip packages after Conda packages. When an environment contains packages installed with Conda and Pip, do not update the packages, but rather create a new environment and reinstall all of the packages to get the latest compatible versions, while avoiding conflicts. The package pyfastnoisesimd [18] requires Microsoft Visual C++ 14.0 or greater. Download and install the `Microsoft C++ Build Tools` from https://visualstudio.microsoft.com/visual-cpp-build-tools. In the installer, check `Desktop development with C++` and then install.

Open3D A few sections of this book use `Open3D` [9], a library for processing and visualizing 3D data. To avoid conflicts, install this in a dedicated environment. To use this environment in Spyder, find the path to this environment, set this as the path for Spyder's Python Interpreter, and restart Spyder.

```
conda create -n open3d -c conda-forge -c open3d-admin open3d spyder-kernels
conda activate open3d
where python
conda deactivate
conda activate spyder-env
spyder
```

References

[1] [Software] Anaconda, *Miniconda* version 23.5.0, 2023. lic: BSD 3-Clause License. url: https://docs.conda.io/en/latest/miniconda.html, vcs: https://repo.anaconda.com/miniconda.

[2] [Software] Conda Contributors, *Conda: A system-level, binary package and environment manager running on all major operating systems and platforms.* Version 23.5.2, 2023. lic: BSD 3-Clause License. url: https://conda.org/, vcs: https://github.com/conda/conda.

[3] [Software] Spyder Project Contributors, *Spyder: the Scientific Python Development Environment* version 5.4.3, 2023. lic: MIT License. url: https://www.spyder-ide.org, vcs: https://github.com/spyder-ide/spyder.

[4] [Software] NumPy Developers, *NumPy* version 1.25.0, 2023. lic: BSD 3-Clause License. url: https://numpy.org, vcs: https://github.com/numpy/numpy.

[5] [Software] David Mans, *Bitmap+* version 1.6.2, 2023. lic: MIT License. url: https://interopxyz.gitbook.io/bitmap-plus/, vcs: https://github.com/interopxyz/BitmapPlus.

[6] [Software] David Mans, *Color+* version 1.5.0, 2023. lic: MIT License. url: https://interopxyz.gitbook.io/color-plus/, vcs: https://github.com/interopxyz/ColorPlus.

[7] [Software] Keyan Rahimzadeh, Ramon van der Heijden, and Alan Tai, *eleFront* version 5.1.2, 2023. Front. lic: MIT License. url: https://elefront.io/, vcs: https://www.food4rhino.com/en/app/elefront.

[8] [Software] SciPy Developers, *SciPy* version 1.11.1, 2023. lic: BSD 3-Clause License. url: https://scipy.org, vcs: https://github.com/scipy/scipy.

[9] [Software] Open3D Development Team, *Open3D: A Modern Library for 3D Data Processing* version 0.17, 2023. lic: MIT License. url: http://www.open3d.org/, vcs: https://github.com/isl-org/Open3D.

[10] [Software] Visose, *Robots* version 1.6.5, 2023. lic: MIT License. url: https://www.food4rhino.com/en/app/robots, vcs: https://github.com/visose/robots.

[11] [Software] Arend van Waart, *Voxeltools: Lightweight voxel geometry and Boolean voxel operations for Grasshopper* version 1.03, 2023. lic: MIT License. url: https://www.food4rhino.com/en/app/voxeltools, vcs: https://github.com/arendvw/voxeltools.

[12] [Software] Association for Robots in Architecture, *KUKA|prc: Parametric Robot Control* 2022. url: https://robotsinarchitecture.org/, vcs: https://www.food4rhino.com/en/app/kukaprc-parametric-robot-control-grasshopper.

[13] [Software] Philip Belesky, *Caribou* version 0.13.5, 2022. lic: GNU Lesser General Public License v3.0. url: https://caribou.philipbelesky.com/, vcs: https://github.com/philipbelesky/Caribou/.

[14] [Software] Landau Design+Technology, *Land Kit* version 1.2, 2022. url: https://www.landkit.design/, vcs: https://www.food4rhino.com/en/app/land-kit.

[15] [Software] Ryan Oenning, *Dendro* version 1.0.0, 2022. lic: Mozilla Public License 2.0. url: https://www.food4rhino.com/en/app/dendro, vcs: https://github.com/ryein/dendro.

[16] [Software] Brian Washburn, *Heron* version 0.4.1, 2022. lic: MIT License. vcs: https://github.com/blueherongis/Heron.

[17] [Software] Philip Belesky, *Groundhog* version 0.13.0b, 2021. lic: GNU General Public License v3. url: https://groundhog.la/, vcs: https://github.com/philipbelesky/groundhog.

[18] [Software] Robert McLeod, *PyFastNoiseSIMD* version 0.4.2, 2021. lic: BSD 3-Clause License. url: http://pyfastnoisesimd.readthedocs.io/, vcs: https://github.com/robbmcleod/pyfastnoisesimd.

[19] [Software] Petras Vestartas and Andrea Settimi, *Cockroach: A plug-in for Rhino point cloud post-processing and meshing* version 1.0.0, 2021. EPFL Laboratory for Timber Constructions. lic: GNU Lesser General Public License v3.0. url: https://www.food4rhino.com/en/app/cockroach, vcs: https://github.com/9and3/Cockroach.

[20] [Software] weArk, *SnappingGecko* version 1.1.2, 2021. lic: CC-BY-SA 4.0. url: https://www.we-ark.fr/, vcs: https://www.food4rhino.com/en/app/snappinggecko.

[21] [Software] Jose Luis García del Castillo y López, *Machina* version 0.8.12, 2020. lic: MIT License. url: https://www.food4rhino.com/en/app/machina, vcs: https://github.com/RobotExMachina.

[22] Will Pearson. *Package Manager Guides*. Robert McNeel & Associates, 2020. url: https://developer.rhino3d.com/guides/yak.

[23] [Software] Ilmar Hurkxkens and Mathias Bernhard, *Docofossor* version 0.904, 2019. lic: MIT License. vcs: https://github.com/dbt-ethz/docofossor.

[24] [Software] Cameron Newnham, *Tarsier* version 1.0.1.0, 2019. lic: GNU General Public License. url: https://www.food4rhino.com/en/app/tarsier, vcs: https://bitbucket.org/camnewnham/tarsier.

[25] [Software] Bison Development Team, *Bison* version 0.9, 2018. url: https://www.food4rhino.com/en/app/bison.

[26] [Software] Henrik Leander Evers and Mateusz Zwierzycki, *Volvox* version 0.3.0.0, 2016. Centre for Information Technology and Architecture. url: https://www.food4rhino.com/en/app/volvox, vcs: https://github.com/CITA-cph/Volvox/tree/master.

[27] [Software] Timothy Logan, *Elk* version 0.3.1, 2013. url: https://www.food4rhino.com/en/app/elk.

[28] [Software] Giulio Piacentino, *4D Noise* version 1, 2011. McNeel Europe. vcs: https://www.food4rhino.com/en/app/4d-noise.

B Color Gradients

Color is critically important for data visualization in both science and design. In landscape architecture and the spatial sciences, color gradients are used in graphics to represent phenomena such as topography, diversity, temperature, air quality, rainfall, flooding, erosion, and fire. When color gradients are used to represent these sorts of numerical data, the gradients should be accurate, legible, intuitive, and ordered. They should clearly represent the full range of data without distortion. A good color map should be perceptually uniform and ordered, accessible for those with color vision deficiency, and legible in grayscale. A perceptually uniform color map has a uniform, linear gradient of perceived lightness and saturation. The colors in such gradients appear equidistant. Because the difference between colors is proportional to the difference between data values, these color gradients can be read accurately [9, 7].

Poorly designed and chosen color maps, however, misrepresent data and exclude readers with color vision deficiencies. Many common color maps – such as `jet`, which has a rainbow gradient across the full spectrum from red to green to blue – have perceptually uneven gradients that distort data, lose data in grayscale, and are inaccessible [7]. Uneven color maps exaggerate some changes in the data, while hiding others. When exaggerations are extreme, they can make continuous data appear discontinuous. With a topographic map, for example, `jet` can make gentle slopes appear as sheer cliffs, while ridges, peaks, and crests appear instead as plateaus (Figure B.3). Colors, therefore, should be carefully chosen to represent data accurately.

Collections of perceptually uniform color maps include `ColorBrewer` [11, 13], `Matplotlib` [3, 12], `Seaborn` [4, 6], `cmocean` [9], `ColorCET` [10], `Scientific Colours` [1, 7], and `CMasher` [5, 8]. These collections include sequential, diverging, and cyclical classes of color maps that can be used to represent continuous, discrete, and categorical data. Sequential color maps represent ordered sequences of data. `viridis` from `Matplotlib` is an example of a perceptually uniform, sequential color map. Diverging color maps represent data that diverge from a critical central value. `delta` from `cmocean`, for example, can be used to intuitively represent how bathymetry and topography diverge from mean sea level. Multi-sequential color maps like `oleron` from `Scientific Colours` can also be used to represent data that have a key, central value. Cyclical color maps are used for data with no beginning or end. These color maps begin and end with the same hue, so they can be smoothly cycled through. `roma0` is a beautiful example from `Scientific Colours`. Some of these maps can be used to intuitively represent themes. Terrain can be represented intuitively with `Matplotlib`'s `cubehelix` or `Scientific Colour`'s `batlow` maps. Ecosystems can be represented with `CMasher`'s `swamp`, `jungle`, `rainforest`, `savanna`, or `seaweed` maps.

Figure B.1: Classes of color maps

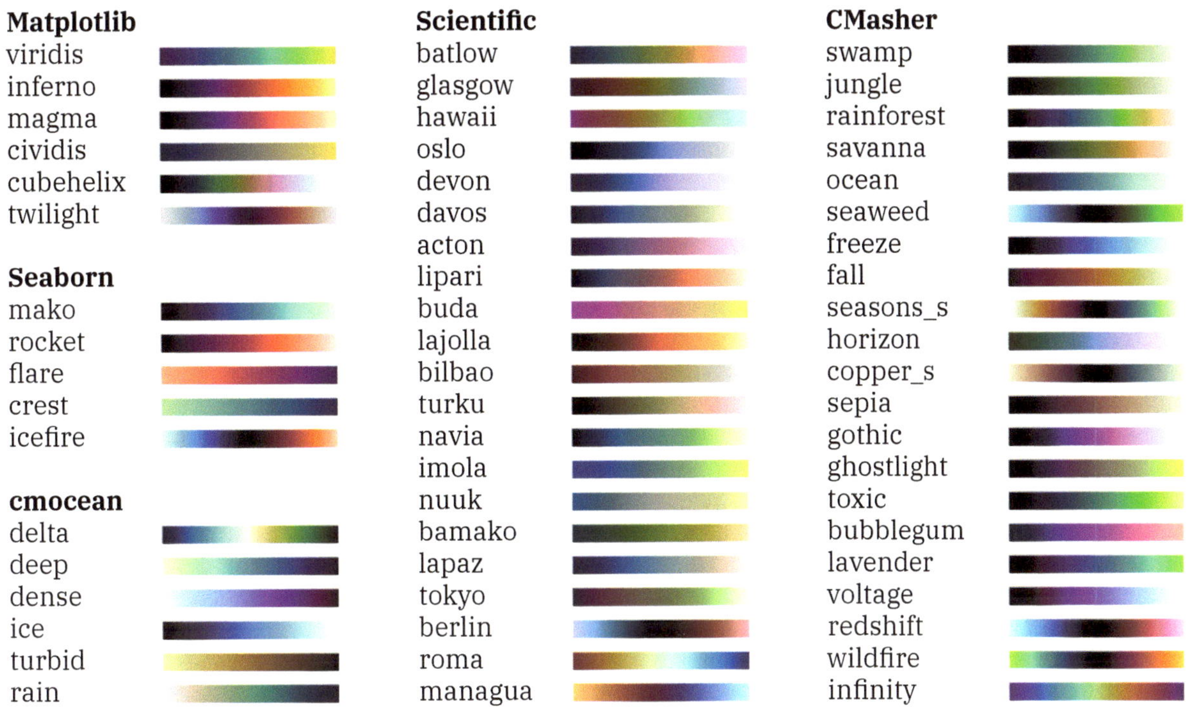

Figure B.2: Perceptually uniform color maps

Code: `color-gradients.py`

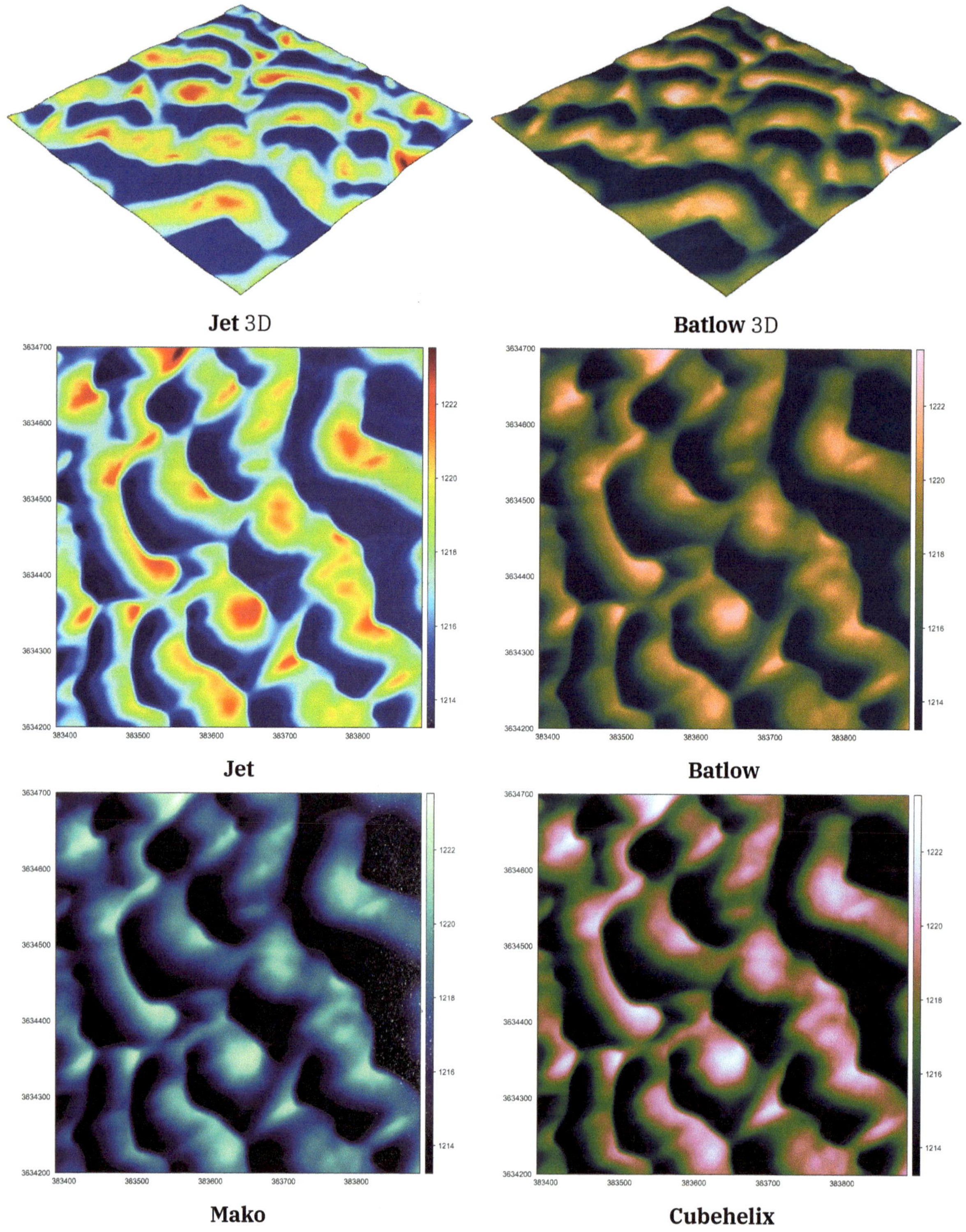

Figure B.3: Color map comparison

B.1 Grasshopper

B.1.1 Custom Gradients

The Grasshopper definition `color-maps.gh` contains a small library of perceptually uniform color maps. Sequential color maps include viridis, inferno, magma, plasma, cividis, mako, and cubehelix, while cyclical color maps include twilight and icefire. To use these color maps, simply copy and paste them into the definition you are developing. See `color-gradient.gh` for an example of how to use one of these color gradients.

Code: `color-maps.gh` & `color-gradient.gh`

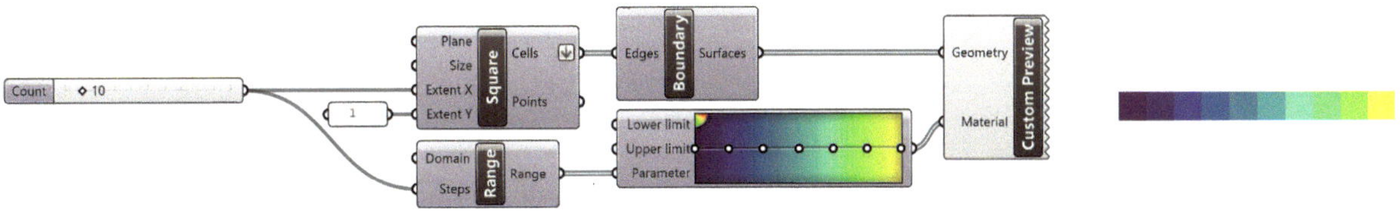

B.1.2 Color+

The `Color+` plugin [2] has an extensive collection of tools for working with color. Use its `Color Palettes` component to select a color map from its library. To use a sequential, perceptually uniform color map such as `Viridis`, select the `*D3 Seq. Spectrum` collection and pick a map from the sub-collections. Then use its `Construct Gradient` component to generate a gradient from this color map. Set the sample parameter to a `Range` of numbers with a domain from `0` to `1`. The number of steps in the range will determine the number of colors in the gradient.

Code: `color-plus.gh`

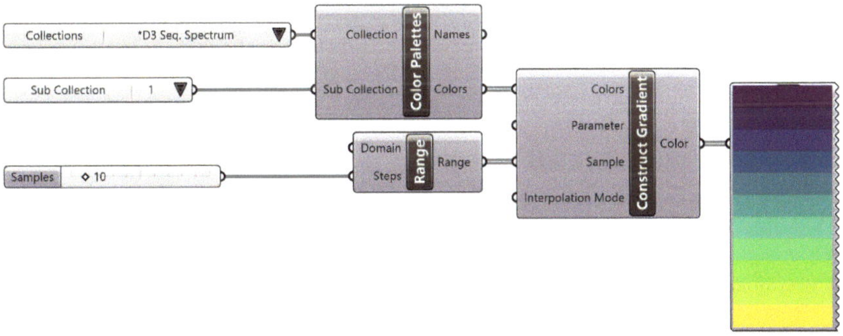

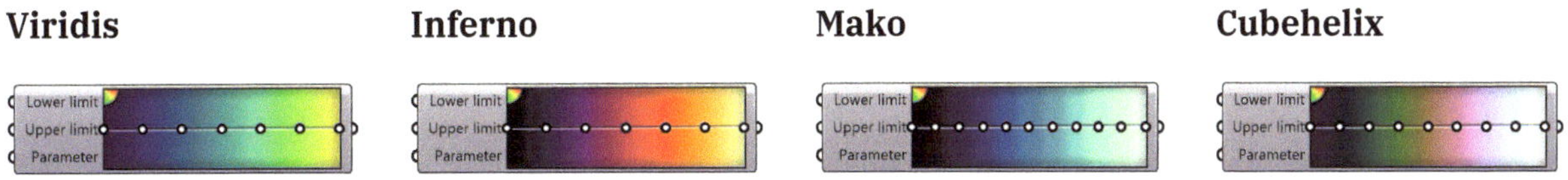

Figure B.4: Custom gradients

B.2 Python

B.2.1 Matplotlib

To use one of `Matplotlib`'s color maps, first import the library and then call `matplotlib.colormaps`.

Code: `matplotlib-gradient.py`

```
# import library
import matplotlib as mpl

# set color map
gradient = mpl.colormaps['viridis']
```

Color maps from other libraries including `Seaborn`, `cmocean`, `Scientific`, and `CMasher` can be accessed via `Matplotlib`. Simply import both libraries and then use `matplotlib.colormaps` to set a color map. For `cmocean` prepend the prefix `cmo` before the name of the color map. For `Scientific` prepend the prefix `cmc` before the name of the color map. For `CMasher` prepend the prefix `cmr` before the name of the color map.

```
import matplotlib as mpl
import cmocean.cm as cmo
mpl.colormaps['cmo.dense']
```

```
import matplotlib as mpl
import cmcrameri.cm as cmc
mpl.colormaps['cmc.roma']
```

```
import matplotlib as mpl
import cmasher as cmr
mpl.colormaps['cmr.wildfire']
```

B.2.2 Seaborn

To use one of `Seaborn`'s color maps, first import the library and then call the function `seaborn.color_palette`.

Code: `seaborn-gradient.py`

```
# import library
import seaborn as sns

# set color map
gradient = sns.color_palette('icefire', as_cmap=True)
```

B.2.3 cmocean

To use one of `cmocean`'s color maps, first import the library and then call `cmocean.cm`.

Code: `cmocean-gradient.py`

```
# import library
import cmocean

# set color map
gradient = cmocean.cm.delta
```

B.2.4 Scientific Colours

To use one of the `Scientific` color maps, first import the library and then call `cmcrameri.cm`.

Code: `cmcrameri-gradient.py`

```
# import library
import cmcrameri.cm as cmc

# set color map
gradient = cmc.managua
```

B.2.5 CMasher

To use one of `CMasher`'s color maps, first import the library and then call `cmasher`. Plot the color map with the function `cmasher.view_cmap`.

Code: `cmasher-gradient.py`

```
# import library
import cmasher as cmr

# set color map
gradient = cmr.gothic

# plot color map
cmr.view_cmap(gradient)
```

References

[1] [Software] Fabio Crameri, *Scientific colour maps* version 8.0.0, June 2023. doi: 10.5281/zenodo.8035877, url: https://doi.org/10.5281/zenodo.8035877.

[2] [Software] David Mans, *Color+* version 1.5.0, 2023. lic: MIT License. url: https://interopxyz.gitbook.io/color-plus/, vcs: https://github.com/interopxyz/ColorPlus.

[3] [Software] Matplotlib Development Team, *Matplotlib* version 3.7.2, 2023. lic: Matplotlib License. url: https://matplotlib.org, vcs: https://github.com/matplotlib/matplotlib.

[4] [Software] Michael Waskom, *Seaborn: statistical data visualization* version 0.12.2, 2022. lic: BSD 3-Clause License. url: https://seaborn.pydata.org, vcs: https://github.com/mwaskom/seaborn.

[5] [Software] Ellert van der Velden, *CMasher: Scientific colormaps for making accessible, informative and cmashing plots* version 1.6.0, 2021. lic: BSD-3-Clause License. url: https://cmasher.readthedocs.io/, vcs: https://github.com/1313e/CMasher.

[6] Michael L. Waskom. "seaborn: statistical data visualization". In: *Journal of Open Source Software* 6.60 (2021), page 3021. doi: 10.21105/joss.03021.

[7] Fabio Crameri, Grace E. Shephard, and Philip J. Heron. "The misuse of colour in science communication". In: *Nature Communications* 11.1 (2020), page 5444. issn: 2041-1723. doi: 10.1038/s41467-020-19160-7.

[8] Ellert van der Velden. "CMasher: Scientific colormaps for making accessible, informative and 'cmashing' plots". In: *Journal of Open Source Software* 5.46 (2020), page 2004. doi: 10.21105/joss.02004.

[9] Kristen M. Thyng, Chad A. Greene, Robert D. Hetland, Heather M. Zimmerle, and Steven F. DiMarco. "True Colors of Oceanography: Guidelines for Effective and Accurate Colormap Selection". In: *Oceanography* (Sept. 2016). url: https://doi.org/10.5670/oceanog.2016.66.

[10] Peter Kovesi. *Good Colour Maps: How to Design Them.* 2015. doi: 10.48550/arXiv.1509.03700. arXiv: 1509.03700 [cs.GR].

[11] [Software] Cynthia A. Brewer and Mark Harrower, *ColorBrewer 2.0: Color Advice for Cartography* version 2.0, 2013. The Pennsylvania State University. lic: Apache-2.0 License. url: https://colorbrewer2.org/, vcs: https://github.com/axismaps/colorbrewer/.

[12] John D. Hunter. "Matplotlib: A 2D Graphics Environment". In: *Computing in Science and Engineering.* 9.3 (June 2007), pages 90–95. issn: 1521-9615. doi: 10.1109/MCSE.2007.55.

[13] Mark Harrower and Cynthia A. Brewer. "ColorBrewer.org: An Online Tool for Selecting Colour Schemes for Maps". In: *The Cartographic Journal* 40 (2003), pages 27–37. doi: 10.1179/000870403235002042.

C File Paths

C.1 Paths

When working with data, it is essential to know where it is stored in order to access it. A file path is where your data is stored on your computer's file system. File paths differ between Windows and Unix-based operating systems like Mac and Linux. Windows, for example, uses backslashes \ as separators in file paths, while Mac and Linux use slashes /. File paths can be specified as absolute or relative paths. An absolute path point specifies the exact location of the file like:

```
C:\computational-design-dataset\02-visual-programming\fundamentals.gh
```

A relative path specifies the location of a file relative to the current working directory. Scripts that use absolute paths are brittle. When a script with an absolute path is run on a different computer with a different file structure, it is unlikely that the path will be found. If data and scripts are stored together, then relative paths can help make scripts more robust. When developing scripts that work with data, carefully consider how to handle file paths.

The code and data in the `dataset` for this book are organized by chapter: Computational Design Dataset > Chapter > File. While the snippets of Python code in the book just specify filenames rather than file paths for the sake of brevity, the scripts themselves use relative paths.

Dataset: `computational-design-dataset`

C.2 Grasshopper

With Grasshopper, an easy approach is to set an absolute file path with a `File Path` component. Simply add the component, right click to select a new file location, and then browse to the file. When a Grasshopper definition using a `File Path` component is moved to another computer, however, the path will likely need to be set again.

C.3 Python

Absolute Paths In Python, backslashes are reserved escape characters, so Windows' file paths cannot simply be specified with strings. They can, however, be specified with raw strings. To write a raw string, simply prepend an `r` before the string.

```
r"C:\computational-design-dataset\c-file-paths\file.txt"
```

Relative Paths If your data is stored in the same directory as your code, then it is easy to specify a robust, relative file path in Python using the `pathlib` module. As long as your data stays in the same location as your code, your script will work as is when it is moved to another computer.

Create a Python script and an empty text file named `file.txt` in the same directory on your computer. In the Python script, first import the modules `pathlib` and `os`. Assign a string with the file name to the variable `filename`. Find the directory containing this Python script with `pathlib.Path(__file__).parent.resolve()`. Then join the path and the file name with the function `os.path.join`. To test the result, print the full path file name.

Code: `relative-path.py`

```
# import modules
import pathlib
import os

# set path
filename = 'file.txt'
datapath = pathlib.Path(__file__).parent.resolve()
data = os.path.join(datapath, filename)
print(data)
```

```
C:\computational-design-dataset\c-file-paths\file.txt
```

C.4 Rhino Scripting

Relative paths can easily be implemented in Grasshopper with a little bit of Python. Start a new Grasshopper definition and add a `Script` component to the canvas. Double click on the script component to open the `Script Editor`. Create a new Python 3 script and save it in the same directory as `file.txt`. In the script, begin by importing the module `os`. Find the path to the current Grasshopper document with `ghdoc.Path` and then find the path of the directory containing the document with `os.path.dirname`. Join the path to the directory with the file name `file.txt` using the function `os.path.join`. Save and close the Python script. Set a panel with the name of the file as an input to `Script` component and set the output to a `File Path` parameter. Make sure the component's parameters and the script's variables match exactly. Right click on the middle of the component to rename it.

Code: `rhino-relative-path.py` & `relative-path.gh` & `grasshopper-relative-path.py`

```
# import modules
import os

# set path
datapath = os.path.dirname(ghdoc.Path)
Path = os.path.join(datapath, Filename)
print(Path)
```

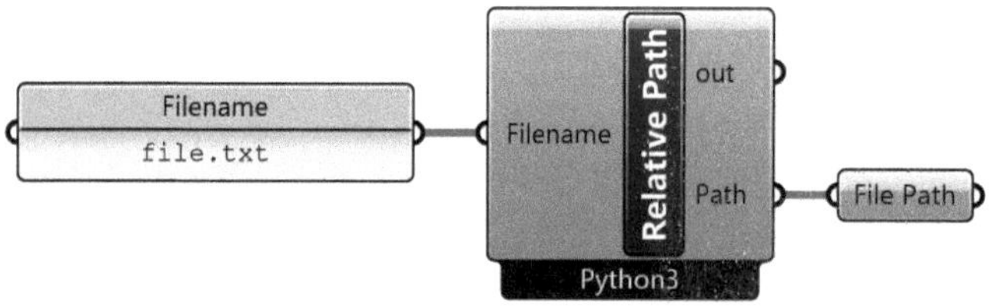

D Code and Data

D.1 Dataset

The Computational Design Dataset contains all of the code and data for this book organized by chapter. It is permanently archived on Zenodo, an online repository. Download the dataset from Zenodo and extract the `zip` archive.

Dataset: `https://doi.org/10.5281/zenodo.8191264`

D.2 Code

The code for this book is also hosted on GitHub.

Code: `https://github.com/baharmon/computational-design-library`

D.3 Cloud Forest

The Cloud Forest library of laser-scanned plants is hosted on our server `xyz.cct.lsu.edu` with interactive, online point cloud visualizations. It is also permanently archived on Zenodo.

Server: `https://xyz.cct.lsu.edu`

Dataset: `https://doi.org/10.5281/zenodo.8194066`

Credits

Engine	XeLaTeX
Typeface	IBM Plex XITS Math
Software	Rhinoceros Python Miniconda Spyder
Data	Open Topography New York City New York State
Images	Benedikt Groß, *Avena+ Test Bed*, 2013 Jared S. Tarbell, *Infinite Regeneration*, 2021 Ryoji Ikeda, *test pattern [live set]*, 2010, photograph by Nacho Pintos Ryoji Ikeda, *data.path*, 2013, photograph by r2hox Davide Quayola, *Pleasant Places*, 2015 Davide Quayola, *Remains: Vallée de Joux*, 2018 dECOi Architects, *One Main*, 2011 MAEID, *Pahoehoe Beauty*, 2018 MAEID, *Magic Queen*, 2021 MAEID, *Sylva*, 2021, photograph by Zita Oberwalder Dominic Jud, Ilmar Hurkxkens, & Marco Hutter, *Robotic Embankment*, 2020

Index